Maritime C ▌▋▌▊▌▌▋▊▌▋ ns

70093357

Maritime Cargo Operations presents the core concepts of cargo work for marine engineering students and cadets. It is built around the essential principles of the maritime profession and is a valuable guide to a broad range of key subject areas in the safe carriage, handling, stowage and securing of cargo, and cargo watches in port. It contributes to a sound understanding of cargo operations for a future career in the profession, as well as offering a general overview for deck officers.

- Gives an overview of the key areas in cargo operations work.
- Includes structured Learning Outcomes and self-test questions for each subject area to assist readers in evaluating their understanding.

This book suits merchant navy cadets at Higher National Certificate (HNC), and Higher National Diploma (HND), and foundation degree level in both the deck and engineering branches, and also serves as a general reference for maritime professionals.

Maritime Cargo Operations

Alexander Arnfinn Olsen

Routledge
Taylor & Francis Group

LONDON AND NEW YORK

Designed cover image: © Shutterstock image 1260654667

First published 2023
by Routledge
4 Park Square, Milton Park, Abingdon, Oxon OX14 4RN

and by Routledge
605 Third Avenue, New York, NY 10158

Routledge is an imprint of the Taylor & Francis Group, an informa business

© 2023 Alexander Arnfinn Olsen

British Library Cataloguing-in-Publication Data
A catalogue record for this book is available from the British Library

ISBN: 978-1-032-40697-8 (hbk)
ISBN: 978-1-032-40696-1 (pbk)
ISBN: 978-1-003-35433-8 (ebk)

DOI: 10.1201/9781003354338

Typeset in Sabon
by Apex CoVantage, LLC

Contents

Preface

This book has been written to encompass three main parts over 15 subject areas. A Learning Outcome is included at the beginning of each part to enable the reader to know what is covered and achieved by the end of each part. There are self-test questions at the end of each part to assist the reader to evaluate their understanding of the subject matter they have just covered. The materials produced here are intended as a guide only. Readers in general, and students in particular, are advised to obtain a copy of the recommended text and references for detailed study to achieve a broader understanding of each of the subjects covered. While studying each subject area, the reader should constantly refer to the references for that particular topic.

RECOMMENDED FURTHER READING

House, D. J., *Cargo Work: For Maritime Operations*, Seventh Edition, Elsevier, 2005

Isbester, J., *Bulk Carrier Practice*, The Nautical Institute, 1993

Roberts, P., *Watchkeeping Safety and Cargo Management in Port*, The Nautical Institute, 1995

International Safety Guide for Oil Tankers and Terminals – OCIMF, Sixth Edition, 2020

Thomas, Agnew et al., *Thomas' Stowage: The Properties and Stowage of Cargoes*, Brown, Son & Ferguson Ltd., 1986

RECOMMENDED REFERENCES

Australian Code of Practice for Ship-Helicopter Transfers

Australian Manual of Safe Loading, Ocean Transport and Discharge Practices for Dry Bulk Commodities

Bulk Carriers: Guidance and Information to Shipowners and Operators

Code of Practice for the Safe Loading and Unloading of Bulk Carriers
Code of Safe Practice for Cargo Stowage and Securing
Code of Safe Practice for Ships Carrying Timber Deck Cargoes, 2011
Code of Safe Working Practice for Australian Seafarers (COSWPAS)
Code of Safe Working Practices for Merchant Seafarers (COSWP)
Guidelines for Offshore Marine Operations
Guidelines for the Preparation of Garbage Management Plans
Guidelines for the Preparation of the Cargo Securing Manual
ICS Publication Tanker Safety Guide (Liquefied Gas)
International Code for the Safe Carriage of Grain in Bulk
International Code for the Safe Carriage of Packaged Irradiated Nuclear
    Fuel, Plutonium and High-Level Radioactive Wastes On Board Ships
    (INF Code)
International Convention for the Prevention of Pollution by Ships, 1973
International Convention on Standards of Training, Certification and
    Watchkeeping for Seafarers, 1978
International Maritime Dangerous Goods (IMDG) Code
International Maritime Solid Bulk Cargoes (IMSBC) Code
International Safety Guide for Oil Tankers and Terminals (ISGOTT)
Liquefied Gas Handling Principles on Ships and in Terminals
Marine Navigation Act, 2013
Marine Order Part 28, 32, 33, 34, 41, 42 and 43
Merchant Shipping (Carriage of Cargoes) Regulations, 1999
Navigation Act, 1912 (Australia):

Division 9	Containers
Division 10	Dangerous Goods, Livestock, Grain Deck and Other Cargoes
Division 12	Ships Carrying or Using Oil
Division 12	A Ships Carrying Noxious Liquid Substances in Bulk

Offshore Support Vessel Code of Safe Working Practice (OSV Code)
Provision and Use of Work Equipment Regulations, 1998 (PUWER)
Safe Loading and Unloading of Bulk Carriers, 2003
Work at Height Regulations, 2005

Acknowledgements

The author is indebted to the Maritime and Coastguard Agency, UK, and the Australian Maritime Safety Authority (AMSA) for allowing the reproduction of the regulations contained herein. The author also wishes to acknowledge the Port of Helsingborg, Sweden, for their permission to reproduce the Ship-Shore Safety Checklist and Guidelines. Further acknowledgements, and personal thanks, to Tony Moore and Aimee Wragg at Routledge, Spandana P.B. at Apex CoVantage, and Fidaa Karkori, whose patience and support in this endeavour has not gone unappreciated.

Figures

Tables

Key Terms

INTERNATIONAL TRADE TERMS

EXW	Ex-works
FOT/R	Free on truck or rail (container)
FOT/R	Free on truck or rail (share a container)
FAS	Free alongside ship
FOB	Free onboard
FOB &	Free onboard and stowed
C & F	Cost and freight
CIF	Cost, insurance, and freight
CIF &	Cost, insurance, freight, and landed
Ex-Quay	Duty paid
EBS	Ex-bond store
FIS	Free into store

Standard Shipping Terms

APCA	Australian port charge additional
BAF	Bunker adjustment factor
B/B	Break bulk
B/L	Bill of Lading
CAF	Currency adjustment factor
CBM	Cubic metre
D/O	Delivery order
FCL	Full container load
H/BL	House Bill of Lading
M	Measurement
M/BL	Master Bill of Lading
M/T	Metric tonne
RORO	Roll-on Roll-off
SCA	Sea cargo automation
THC	Terminal-handling charge
TRC	Terminal receival
W	Weight

Charter Party Laytime Terms

All time saved	The time saved to the ship from the completion of loading/discharging to the expiry of the laytime, including periods excepted from the laytime
All working time saved or all laytime saved	The time saved to the ship from the completion of loading/discharging to the expiry of the laytime excluding any notice time and periods excepted from the laytime
As fast as the vessel can receive/ deliver	The laytime is a period of time to be calculated by reference to the maximum rate at which the ship in full working order is capable of loading/discharging the cargo
ATUTC	All time used to count
CHOPT	Charterers option (often refers to a discharge port)
Demurrage	The money payable to the owner for delay for which the owner is not responsible in loading and/or discharging after the laytime has expired
Despatch money or despatched	The money payable by the owner if the ship completes loading or discharging before the laytime has expired
Excepted	The specified days do not count as laytime even if loading or discharging is done on them
FIOT	Free in, out, and trimmed. Free to the ship, with the charterer organising and paying for the stevedoring
Holiday	A day of the week or part(s) thereof on which the notice is given and the day on which the notice expires are not included in the notice period
In writing	In relation to a notice of readiness, a notice visibly expressed in any mode of reproducing words and includes cable, telegram, telex, fax (email requires the consent of all parties prior to signing of contract)

Laytime	The period of time agreed between the parties during which the owner will make and keep the ship available for loading/discharging without additional payment to the freight
Liner terms	Applicable to a liner trade where an owner has his ships regularly calling at his arranged berths and thus can offer to load and discharge incorporating stevedoring into the freight rate
Notice of readiness	Notice to the charter, shipper, receiver, or other person as required by the charter that the ship has arrived at the port or berth as the case may be and is ready to load/discharge
On demurrage	The laytime has expired. Unless the charter party expressly provides to the contrary, the time on demurrage will not be subject to the laytime exceptions
Per hatch per day	Laytime is to be calculated by multiplying the agreed daily rate per hatch of loading/discharging the cargo by the number of the ship's hatches and dividing the quantity of cargo by the resultant sum.

Thus:

Laytime = quantity of cargo = days

Daily rate × number of hatches

A hatch that is capable of being worked by two gangs simultaneously shall be counted as two hatches

Per working hatch per day or per workable hatch per day	Laytime is to be calculated by dividing the quantity of cargo in the hold with the largest quantity by the result of multiplying the agreed daily rate per working or workable hatch by the number of hatches serving that hold.

Thus:

Laytime = largest quantity in one hold = days

Daily rate per hatch × number of hatches serving that hold

A hatch that is capable of being worked by two gangs simultaneously shall be counted as two hatches

Reachable on arrival or always accessible	The charterer undertakes that when the ship arrives at the port, there will be a loading and discharging berth for her to which she can proceed without delay

Reversible	An option given to the charterer to add together the time allowed for loading and discharging. Where the option is exercised, the effect is the same as a total being specified to cover both operations
Running days or consecutive days	Days which follow one immediately after the other
Safe berth	A berth that, during the relevant period of time, the ship can reach, enter, remain at, and depart from without, in the absence of some abnormal occurrence, being exposed to danger which cannot be avoided by good navigation and seamanship
Safe port	A port that, during the relevant period of time, the ship can reach, enter, remain at, and depart from without, in the absence of some abnormal occurrence, being exposed to danger which cannot be avoided by good navigation and seamanship
SHINC	Saturday or Sunday holidays included (in time)
Time lost waiting for berth to count as loading discharging time or as laytime	If the main reason why a notice of readiness cannot be given is that there is no loading/discharging berth available to the ship, the laytime will continue to run. Where the ship continues to wait for a birth to become available, the laytime will continue until such time as a berth becomes available. The laytime exceptions apply to the waiting time as if the ship was at the loading/discharging berth, provided the ship is not already on demurrage. When the waiting time ends, time ceases to count and restarts when the ship reaches the loading/discharging berth subject to the giving of a notice of readiness if one is required by the charter party, and to any notice time if provided for in the charter party, unless the ship is by then on demurrage
To average	Separate calculations are to be made for loading and discharging, and any time saved in one operation is to be set against and any excess time used in the other
Unless used	If work is carried out during the excepted days, the actual hours of work only counts at laytime
Weather permitting	Time during which weather prevents working shall not count as laytime

Weather working day	A working day or part of a working day during which it is or, if the vessel is still waiting for her turn, it would be possible to load/discharge the cargo without interference due to the weather. If such interference occurs (or would have occurred if work had been in progress), they shall be excluded from the laytime a period calculated by reference to the ratio which would have or could have been worked but for that interference
Weather working day of 24 consecutive hours	A working day or part of a working day or 24 hours during which it is or, if the ship is still waiting for her turn, it would be possible to load/discharge the cargo without interference due to the weather
Whether in berth *or* not or berth *no* berth	If the location named for loading/discharging is a berth and if the berth is not immediately accessible to the ship, a notice of readiness can be given when the ship has arrived at the port in which the berth is situated
Working days	Days of or part(s) thereof which are not expressly excluded from laytime by the charter party, and which are not holidays
1 SB PORT	One safe berth or port

Part 1

Safe Carriage of Cargo

Principles and Practice

LEARNING OUTCOME 1

On completion of Part 1, you should be able to demonstrate knowledge of the principles and practices applicable to the safe carriage of cargo by ships at sea and knowledge of the regulations governing the carriage of cargo and cargo-handling equipment.

ASSESSMENT CRITERIA

1.1 Ship's responsibilities and obligations with respect to the carriage of cargo are stated with reference to the following:
- Make holds fit and safe for the reception, carriage, and preservation of cargo.
- Make the ship seaworthy.
- Properly handle, equip, and supply the ship.

1.2 The following basic concepts of cargo carriage are discussed:
- Compatibility of cargoes.
- Suitability of cargo for the ship.
- Suitability of ship for the cargo.
- Whether the cargo can be safely stowed and secured onboard the ship and transported under all expected conditions during the intended passage.

1.3 The following terms used to indicate 'cargo quantity' are defined:
- Bale volume.
- Grain volume.
- Dead weight.
- Tank volume.
- TEU.
- Lane metres.

1.4 Methods of determining the cargo quantity, including draught survey, are outlined.

DOI: 10.1201/9781003354338-1

1.5 Importance of ballast management is explained, and procedures outlined.

1.6 Equipment and instruments used in cargo quantity measurement are stated.

1.7 Principles and practice of safe handling, stowage, and carriage of following cargoes are explained:
- General cargo.
- Unitised cargo.
- RORO cargo.
- Reefer cargo.
- Solid bulk cargo.
- Liquid cargo.
- Offshore supplies.

1.8 The requirements as per international regulations with respect to the carriage of timber deck cargo are outlined.

1.9 The requirements as per national and international regulations with respect to the carriage of grain cargo are outlined.

1.10 The requirements as per national regulations with respect to the carriage of livestock are outlined.

1.11 The requirements as per Marine Order Part 32 with respect to the cargo-handling equipment are outlined.

Chapter 1

Key Concepts of Cargo Work

1.1 INTRODUCTION

The maritime and logistics industry is the backbone of global trade. Each year billions of tonnes of cargo are shipped around the world over oceans, across lakes, and via waterways. Over the past hundred years or so, the world of cargo operations has evolved considerably from the time when cargo was backbreakingly loaded onto ships by scores of longshoremen. In fact, the conception of unitised cargoes in the form of containers, roll-on roll-off (RORO) cargoes, and palletisation has revolutionised the way the shipping industry works. The role of the stevedore and longshoreman has evolved into an entirely different profession based on technology and ever-improving efficiency. Modern techniques of cargo loading are heavily mechanised with gantry cranes, self-loaders, and mobilisation replacing traditional forms of dock labour. Indeed, the effect of containerisation cannot be underestimated nor overstated as it has had an immeasurable influence on the work of the modern seafarer.

1.2 KEY CONCEPTS OF CARGO WORK

In this chapter, we will examine some of the key ideas and terms that will be used throughout this book. When goods are destined to be transported from one location to another, the shipper and the shipowner enter a legally binding contract so that neither party is disadvantaged in the event the contract is not fully complied with. In the UK, this contract is covered by an Act of Parliament referred to as the *Carriage of Goods by Sea Act, 1992*. This replaced an earlier Act of Parliament dating from 1971 which incorporated the 1931 *Hague-Visby Rules* into English Law. These rules themselves were later amended with the adoption of the *Hamburg Rules* in 1978. In 2009 further amendments were proposed (the *Rotterdam Rules*), though as of 2022 these remain officially unratified. In essence, the *Carriage of Goods by Sea Act, 1992,* sets out the obligations, responsibilities, and liabilities

DOI: 10.1201/9781003354338-2

of the person carrying the cargo (i.e. the carrier, cargo owner, or shipper). Most countries have some form of legislation that mirrors the 1992 Act. In the United States, this is the *Carriage of Goods by Sea Act, 1936*, and in Australia, seaborne trade is overseen by the *Australian Carriage of Goods by Sea Act, 1991*. The carrier is defined as someone who, under a special contract, receives goods for the purpose of shipping them from one place to another. Thus, the carrier is legally responsible for the goods whilst the goods are in their possession. Under the *Hague-Visby Rules*, the term carrier includes anyone engaged as the owner or charterer of a vessel and who enters a contract with the shipper. The shipper is effectively the person or entity who has the right to move the goods from one location to another. The shipper may not necessarily be the owner of the goods; this is likely so when the shipper is acting as an agent on behalf of the goods owner. This role is typically performed by freight forwarders. Once the shipper arranges for the goods to be shipped, a document is drawn up which sets out the agreement between the shipper and the carrier. This document is called the *Bill of Lading*, and whilst there is no formal requirement for Bills of Lading to follow any particular style or format, most follow the pro forma provided by the *Hague-Visby Rules*. Refer to Figure 1.1 for an example of a standard Bill of Lading.

Under the *Hague-Visby Rules* and its subsequent amendments, the shipowner is required both before and when commencing the voyage to (a) make the ship ready and seaworthy; (b) properly man, equip, and supply the ship; and (c) make the holds, refrigerating, and cooling chambers, and all other parts of the ship in which the goods are to be carried, fit and safe for their reception, carriage, and preservation for the duration of the passage. These requirements are further stipulated under the *Carriage of Goods By Sea Act, 1992*, which states that a ship is deemed to be seaworthy only when it is in a fit state as to the condition of its hull and equipment, boilers and machinery, stowage of ballast or cargo, the number and qualifications of crew members including officers, and in every other respect, to encounter the ordinary perils of the voyage contracted. Moreover, the ship's officers must ensure, insofar as is practicable, the vessel is not overloaded or stowed in such a way as to adversely impact or affect the ship's stability. The shipowner must ensure the crew employed onboard the vessel is capable of performing their duties safely and effectively. This includes, but is not necessarily limited to, the proper handling, stowage, securing, carrying, and care for the cargo for as long as the cargo is in the ship's possession. This also means that the equipment onboard must be functional and fit for purpose.

The senior officer onboard – the captain or master – is the shipowner's representative and as such is obligated to comply with the provisions of the *Carriage of Goods by Sea Act, 1992* (for UK-registered vessels or vessels

Carrier Name

NON-NEGOTIABLE COPY

SHIPPER	DOCUMENT NUMBER
	EXPORT REFERENCES
CONSIGNEE	FORWARDING AGENT REFERENCES
	POINT AND COUNTRY OF ORIGIN (FOR THE MERCHANT'S REFERENCE ONLY) FL-FLORIDA
NOTIFY PARTY	DOMESTIC ROUTING/EXPORT INSTRUCTIONS (FOR THE MERCHANT'S REFERENCE ONLY).
PRECARRIAGE BY * / PLACE OF RECEIPT *	
VESSEL / VOYAGE NO / PORT OF LOADING	FINAL DESTINATION (FOR THE MERCHANT'S REFERENCE ONLY)
PORT OF DISCHARGE / PLACE OF DELIVERY	

PARTICULARS FURNISHED BY THE MERCHANT — PAGE 1

Container No., Seal No.; Marks & No's	No. of Containers or Packages	H * * M	Kind of Packages; Description of Goods	Gross Weight/Code	Measurement/Code

TOTAL NUMBER OF CONTAINERS OR PACKAGES (IN WORDS)

ITEM NUMBER	WEIGHT	MEASUREMENT	CUR.	RATE	BASIS	P/C	PREPAID	COLLECT

LADEN ON BOARD THE VESSEL		Number of Original Bills of Lading	Place of Bills of Lading Issue	Prepaid at	Payable at
DATE	BY	THREE			

(TERMS OF BILL OF LADING CONTINUED FROM BACK HEREOF) ENLARGED VERSION OF BACK CLAUSES IS AVAILABLE UPON REQUEST FROM THE CARRIER. In witness whereof, the undersigned, on behalf of , the Master and the owner of the Vessel, has signed the number of Bill(s) of Lading stated above, all of this tenor and date, one of which being accomplished, the others to stand void.

Dated Month Day Year B/L No.

Declared Valued (U.S. $) ____________

Figure 1.1 Generic combined Bill of Lading.

Source: Author's own. Magellan Maritime Press Ltd.

sailing into and out of UK ports). From the moment the cargo is loaded onto the vessel, it becomes the master's responsibility. These responsibilities include ensuring the cargo is properly handled and stowed, delivered to the port of discharge in the same condition as when it was loaded onboard the vessel. In practicality, this responsibility is delegated to the ship's cargo officers (typically the chief and first officers), who in turn delegate specific tasks to the ship's crew. It is important to recognise that whilst the master may delegate the practical aspects of loading and discharging cargo onto and off the vessel, they remain wholly responsible for the operation in total. In other words, should anything happen to the cargo whilst it is in the ship's custody, it is the master who must answer for it. That is not to say, of course, that any negligence or poor practice on the part of individual crew members will not go unrecognised.

Insofar as we have very briefly covered the main legal framework around the carriage of goods by sea, it is worth discussing for a moment the main types of cargo that ships are required to transport around the world. The types of cargoes that are normally carried by ships can be split into one of four separate categories:

(1) Bulk-dry or liquid cargoes.
(2) Break bulk, also known as general cargo, including cases, pallets, and drums.
(3) Containerised cargo.
(4) Rolling cargo.

To carry these various types of cargo, different types of ships are needed. Some vessels are specialised for the carriage of only one type of cargo, for example, liquefied natural gas (LNG), which is pumped into specially designed holding tanks on gas carriers. Other types of ships are more versatile and can accommodate a variety of cargoes. Container or box ships generally carry containers only, but these can be loaded with all manner of cargo, including vehicles, break bulk cargoes, and even liquefied and gas products in tanks. A unique design of vessel emerged in the 1970s and remained popular until the mid-1990s. The CONRO or container/RORO vessel was a hybrid design which incorporated a large open weather deck for stowing containers and an open plan interior with adjustable tween decks. To facilitate the loading and discharging of rolling stock, CONROs had a large angled stern ramp. Today, CONROs have been replaced largely by box ships and dedicated vehicle carriers, though some continue to remain in service mainly on routes between the United States and Asia. Another type of ship that has fallen out of popularity is the general cargo ship. Until the advent of containerisation in the 1950s and 1960s, general cargo ships were the mainstay of the global shipping industry. These vessels had large open holds with a removable deck plate. Inside the holds anything from grain to coal to livestock could be shuttled from port to port. Like the CONRO, general cargo ships have largely been replaced by specialised types of vessels dedicated to a particular trade or cargo.

Another feature of some ship types is the variety of equipment that is carried onboard. One such defining feature are ship-mounted cranes or gears. Gears are typically found on smaller vessels such as coasters and small container ships, which tend to call at smaller regional ports that lack their own shoreside infrastructure. Alternatively, many bulk carriers also carry their own gantry cranes for loading and discharging cargo, for much the same reason. Other types of specialist vessels, such as offshore construction and drilling vessels, may have their own derricks for loading and unloading pipelines and other associated construction materials and equipment in the field. We will discuss different ship types and cargo loading and discharging equipment in later chapters, but suffice it to say here that ships with gears (geared ships) are less popular as they are more expensive to purchase, maintain, and operate when compared to vessels without gears (gearless ships). Equally as important to cargo-loading and cargo-unloading operations is ship ballasting. Ballasting is quite a complex and technical subject and, as such, falls outside the scope of this book, but it is worth mentioning that all vessels maintain their stability through ballast. Ballast is any form of weight that helps to keep a ship stable in water. This is important for us to know as when ships either load or unload cargo, the stability of the vessel is adversely affected. When cargo is loaded, the ship will become heavier; the opposite is true when cargo is removed. Moreover, ships are not built with equal weight distribution. Most bulk carriers, for example, are extremely long and empty (when unladen). Given the ship's accommodation and engine department are located towards the stern of the vessel, this means there is a natural tendency for the ship to lean or trim backwards. Too much stern trim and the ship will be unsafe to navigate. Alternatively, some vessels have their accommodation situated towards the midship or even at the bow. Offshore vessels usually have large aft weather decks, with the crew accommodation and bridge located forward. This presents similar issues as our bulk carrier, in the sense that offshore vessels are 'bow heavy'. This problem is offset by the addition of ballast. Moreover, as cargo shifts and moves around the hold, for instance in heavy seas, this can present difficulties for the ship's navigation officer to maintain stability. There are many other factors which affect and influence ship stability such as freshwater and salt water, summer and winter loading, currents, ship design and construction, propulsion systems, and so forth. Whereas it is critical for deck officers to have a good understanding of these factors, they do fall outside the scope of this book, and, therefore, the focus here is solely on ballast as a cargo management measure.

All ship operators (i.e. the owner or charterer) must fully consider the type(s) of cargo they may be required to carry before building, buying, or chartering a vessel. The suitability of the vessel must then be assessed for each task it is then contracted to perform. If the vessel is not suitable for a particular type of cargo, then there arises a serious risk of damage to the cargo. This risk can be substantial. It may also be dangerous for the vessel to carry a type of cargo that is unsuited to the vessel's capabilities. Doing so could render the ship unseaworthy, which would nullify the ship's insurance. In the event of an accident or incident, this may also result in significant damage claims and even prosecution in civil court. One aspect that must not be overlooked is the safety of the cargo, and of course, the integrity of the vessel itself. The ability to fully secure cargo onboard must always exist and must never be knowingly compromised. Vessels will normally incorporate securing arrangements for the type of cargo to be carried as part of its standard equipment. This securing equipment must be sufficiently strong and robust to withstand typical oceangoing conditions. Of course, this does not necessarily mean that the ship's equipment will be fail proof. Rather, it is the seafarer who must exercise caution and take reasonable measures to ensure the voyage is performed as safe as possible. To assist the ship's crew in this endeavour, all vessels must carry a *Cargo Securing Manual*, *Rigging Plan* and specific stability data for the vessel. When used in conjunction, these should greatly aid in safe cargo operations.

We have already briefly touched on some of the main types of ships in today's Merchant Navy, but it is worth spending a few moments discussing the various classifications of ships, as this influences the type of cargo carried, their operational areas, and, subsequently, how cargo should be loaded, stowed, and discharged. In the most simplistic sense, ships can be divided into dry and wet types. Dry ships carry cargo that is in solid form. This includes containerised cargoes, steel pipes and rolls, vehicles, construction plant, leisure craft, paper, waste materials, almost anything that is solid in nature. Dry cargo also covers solid break bulk cargo such as grain, coal, and metal ores, though it is worth mentioning these exhibit the same qualities as wet cargo, something we will discuss later. Wet cargo, as the name suggests, is any cargo that is liquefied and not contained in some form of packaged unit. This covers cargoes such as crude oil, bulk chemicals, liquefied natural gas (LNG), and liquefied petroleum gas (LPG). Where liquefied cargoes are transported in modular units, such as in tanks, these are classified as dry cargo as they are transported on dry ships (such as container ships) as opposed to in the ship's cargo tanks. In the 1970s, a rather odd type of vessel was developed which could simultaneously carry dry, liquid, and bulk cargoes. These ships, referred to as ore, bulk, and oil carriers or OBOs, were not particularly successful, and only a handful were built. Most, if not all, have since been retired from service. By most standards, there are 14 classes of vessels: (1) coal carriers, (2) coasters, (3) combi carriers or CONROs, (4) container or box ships, (5) fruit carriers, (6) gas

Figure 1.2 Typical coal carrier – *MV Katagalan Wisdom III.*
Source: Ossewa, CC BY-SA 4.0.

carriers, (7) Lighter Aboard or LASH ships, (8) multipurpose general cargo ships, (9) ore/bulk/oil carriers, (10) refrigerated vessels, (11) roll-on roll-off (RORO) vessels, (12) tankers, (13) timber carriers, and (14) vehicle carriers.

In summary: *Coal carriers*, as the name suggests, are designed to carry coal in bulk on deep-sea routes (Figure 1.2). They can be gearless or carry their own gear. Facilities for monitoring the cargo temperature and gas emissions form an essential part of the basic equipment carried onboard. Special provisions for pumping bilge water are also present to ensure the ship remains stable in all cargo load conditions.

Coasters are a type of all-purpose cargo carrier which operate around coastal regions (Figure 1.3). They are usually fitted with two holds and carry their own cranes or derricks. This enables coasters to visit ports which lack quayside loading and discharging infrastructure. Because coasters tend to be small, they are frequently used to ship cargoes inland via waterways and canals.

Combi carriers or *CONROs* are a specially designed type of vessel which has a flat weather deck for transporting containerised cargoes and a stern ramp for loading vehicles or rolling cargoes (Figure 1.4). Sometimes these vessels are equipped with cranes to facilitate the loading and discharging of heavy indivisible goods. From the mid-1970s to

Figure 1.3 Typical coaster – *MV Ragna*.

Source: Wolfgang Fricke, CC BY 3.0.

Figure 1.4 Typical combi carrier or *CONRO* – *MV Grande Lagos*.

Source: Ra Boe, Magellan Maritime Press Ltd.

Figure 1.5 Typical container vessel – MV Maersk Elba.

Source: Author's own. Magellan Maritime Press Ltd.

the mid-1990s, these types of ships were very popular, though they have since been replaced by standard container or box ships and specialised RORO carriers.

Container vessels form the backbone of the global merchant fleet as they have become increasingly popular with shipowners (Figure 1.5). Unlike combi carriers, container vessels have large and segmented holds which are topped with a sliding or liftable deck plate called a hatch. The largest container ships in operation today are called megaships as they can carry in excess 23,000 twenty-foot equivalent unit (TEU) containers at any given time.

1.2.1 Fruit Carriers

Like refrigerated or reefer vessels (see Figure 1.6), fruit carriers have specially designed cool-air systems that are installed into their holds. These systems help prevent the fruit from overripening during transit. It also helps prevent the fruit from spoiling as the vessel sails from exotic climates to cooler climates. Typical fruits carried include bananas, apples, oranges, and grapes. Most fruit carriers operate on routes between South America/the Caribbean and Europe, Australia/New Zealand and Europe, and South Africa and Europe. Similar routes also exist with the United States.

1.2.2 Gas Carriers

Most gas carriers or gas tankers now use a combination of specialist materials (such as foam and alloy metals) to provide insulation between the gas tanks and the ship's hull (Figures 1.7 and 1.8). The largest types of

Figure 1.6 Typical fruit carrier – *MV Star Leader*.

Source: Hugh Llewelyn, CC BY-SA 2.0.

Figure 1.7 Typical LNG carrier – *MV Arctic Princess*.

Source: Joachim Kohler, CC BY-SA 4.0.

Figure 1.8 Typical LNG carrier – *MV Christophe de Margerie.*
Source: Kremlin, CC BY-SA 4.0.

gas carriers in operation today employ the Moss Rosenberg system, which consists of large aluminium alloy spheres that contain the liquid gas at a temperature of about –160°C (–256°F). The gas is maintained at steady pressure below 2 bar. The spheres are supported and connected to the hull by an 'equatorial ring' type construction. This consists of a steel alloy ring which is positioned around the circumference of the sphere. The spheres are then insulated by foam and topped with an aluminium cover. Some smaller gas carriers use semi-pressurised type steel tanks which are insulated with foam to protect the tank from the ship's hull.

1.2.3 LASH Vessels

LASH stands for lighter aboard ship. These types of vessels are designed to carry several fully laden lighters or unpropelled barges on deck. The cargo is discharged by the ship's own handling gear when the vessel is at anchorage.

1.2.4 Multipurpose General Cargo Ships

Multipurpose general cargo ships are vessels suitable for worldwide trading in general cargoes, dry bulk, long steel products, grain cargoes, and containers (Figure 1.9). These vessels are most commonly geared in that

Figure 1.9 Typical general cargo ship – *MV Gatun*.

Source: Author's own, Magellan Maritime Press Ltd.

they carry their own cranes or derricks. Ship-mounted cranes and derricks are formally referred to as gears. These types of vessels are less popular today than was the case up to the mid-1990s, though they remain common in some parts of the world such as the Asian subcontinent and in South America, where ports are either very small or have poor quayside facilities.

Ore/bulk/oil carriers (OBO) are multipurpose bulk carriers designed to switch between bulk shipments of oil, bulk grain, fertilisers, and ore trades (Figure 1.10). They are normally gearless, relying on shore facilities for their cargo operations. The average size of OBOs is 270,000 gross tonnes deadweight putting them squarely in the same league as oil tankers.

1.2.5 Refrigerated Vessels

Like the fruit carriers discussed earlier, refrigerated vessels or reefer ships carry chilled or frozen cargo such as eggs, butter, and meat. They have large insulated holds with refrigerating machinery to maintain the cargo in a stable and chilled condition.

Figure 1.10 Typical OBO carrier – *MV Naess Crusader.*

Source: Hugh Venables, Magellan Maritime Press Ltd.

1.2.6 Roll-on Roll-off Vessels

RORO vessels are commonly used as ferries (Figure 1.11). These ships have large open interior decks which are ideal for stowing hundreds of cars and lorries and other forms of rolling stock. All RORO vessels have a stern ramp, which facilitates the ingress and egress of traffic onto and off the vessel. Some ferries also have a bow-mounted ramp, which enables traffic to drive onto and then off the ship in the same direction. This makes manoeuvring traffic onboard the ship much easier and efficient as every vehicle is either facing forwards or backwards. Whereas some ferries are dedicated to short sea routes only, for example, between the UK and France, others traverse longer distances and so need to accommodate passengers in cabins, as opposed to in day rooms. These types of ferries are referred to as ROPAX ferries, as they combine many of the features of cruise ships with cargo-carrying capacity.

1.2.7 Tankers

Tankers are some of the most instantly recognisable types of ships as their size and design are unique amongst the categories of ships we are discussing in this book. Tankers range in purpose from the carriage of crude oils

Figure 1.11 Typical RORO vessel – *MV Clansman*.

Source: Author's own, Magellan Maritime Press Ltd.

to crude and refined products, such as chemicals. They range significantly in size and weight, with the smallest being as little as a few thousand gross tonnes deadweight to the largest, which can exceed 350,000 gross tonnes deadweight (Figures 1.12 and 1.13).

1.2.8 Timber Carriers

Timber carriers are a special type of dry cargo ship designed to carry timber cargo (Figure 1.14). These ships have large hatches to accommodate a variety of different sized timber logs. In many cases, cargo may also be carried on deck. Timber carriers have special load lines as they must abide to more stringent regulations given timber's ability to absorb vast quantities of water, which can adversely affect the ship's stability.

1.2.9 Vehicle Carriers

Last of all, *vehicle carriers* are specially designed vessels which are used for the mass transportation of cars, vans, lorries, and other types of oversized cargoes such as construction equipment. Unlike RORO and ROPAX ships, vehicle carriers do not carry passengers. Instead, they are mostly chartered by vehicle manufacturers to ship new vehicles from the manufacturing plant (or nearest port) to whichever market the vehicles are destined for. There are

Figure 1.12 Typical oil tanker – *MT Safwa.*

Source: Danny Cornelissen, Magellan Maritime Press Ltd.

Figure 1.13 Typical chemical tanker – *MT British Captain.*

Source: Author's own, Magellan Maritime Press Ltd.

Figure 1.14 Typical timber carrier – *MV Star Atlantic.*

Source: Author's own, Magellan Maritime Press Ltd.

Figure 1.15 Typical vehicle (PCC) carrier – *MV Artemis Leader.*

Source: Author's own, Magellan Maritime Press Ltd.

Figure 1.16 Typical vehicle (PCTC) carrier – *MV Topeka*.

Source: Author's own, Magellan Maritime Press Ltd.

two main subclasses of vehicle carrier: the pure car carriers (PCC), which transport only cars, and pure car and truck carriers (PCTC), which transport both cars and larger-sized rolling stock such as lorries (Figures 1.15 and 1.16). The defining difference between PCC and PCTC vessels is in the number, height, and configurations of the internal decks. PCTCs generally have more flexibility in how the internal or tween decks can be configured to work around large and cumbersome cargoes. One design feature that is common to both PCC and PCTCs are the provisions for the extraction of exhaust fumes which are emitted from vehicles as they are driven onto and off the vessel. Like combi vessels, vehicle carriers possess huge stern and occasionally side, mounted ramps.

1.2.10 Others

For the observant reader, it might seem apparent that we have missed out one obvious type of vessel from our list, albeit it did get a very short mention. Cruise ships form an important though relatively niche part of the merchant fleet, and whilst passengers might be considered human cargo in the same sense that livestock carriers transport sheep, cows, and pigs, a

Figure 1.17 Typical cruise ship – *MV Norwegian Joy.*

Source: Georges Desipris, Pexels.

conscious decision was made not to include cruise ships in this book for that same reason (Figure 1.17). For the sake of clarity, however, cruise ships employ a combination of dry and wet cargo management techniques. When seen in isolation, one person has a negligible impact on ship stability. Magnify that effect by an order of 1,000–3,000, and the impact is much more pronounced. In that sense, passengers behave in much the same way as liquid cargo – always on the move, waxing and waning, sometimes coagulating in large groups, other times dispersed. This can make managing cruise ship stability difficult, especially in heavy weather or when passing points of interest (when large numbers of passengers are likely to mass in one area on deck). This can expose the vessel to unpredictable stability. For anyone interested in reading more about vessel stability, and the impact of human cargo, read Barrass and Derrett's *Ship Stability for Masters and Mates*.[1]

That concludes Chapter 1. In the next chapter, we will turn our attention to the process of determining cargo quantity and volumes.

NOTE

1 Barrass, C.B. and Capt. D.R. Derret. 2012. *Ship Stability for Masters and Mates.* Oxford: Butterworth-Heinemann. (ISBN 978-0080970936).

Chapter 2

Determining Cargo Quantity and Volume

As we noted at the beginning of Chapter 1, the shipping and logistics sector is a global industry, which employs hundreds of thousands of people from dozens of countries. To ensure a degree of consistency around the world, English has been adopted as the common and recognised language of the sea. This means there are standardised technical terms which maritime professionals use, both onshore and onboard. Many of these terms are used to express cargo quantities and volumes. In this chapter, we will briefly discuss some of the most common that seafarers are likely to encounter or, indeed, use themselves.

2.1 TERMS USED FOR MEASUREMENT

The following are the most frequently used terms (in alphabetical order) to describe a specific quantity or volume of cargo:

- *Bale capacity:* The total amount of cargo space available for the carriage of bags, bales, pallets, or boxed cargoes and does not include the space between the frames and beams. It is always slightly less than the grain capacity.
- *Broken stowage:* The space that is lost in a ship by cargo of irregular shape or because the hold is an irregular shape and includes the space lost between cargo packages. It is expressed as a percentage, for example, if a cargo has a broken stowage of 5%, this means that an additional 5% of the cargo volume must be allowed for when allocating space for that specific load.
- *Dead weight tonnage:* The amount of cargo, stores, and fuel that a vessel can carry. It is the difference between the summer displacement and the light displacement.
- *Expansion coefficient:* The amount by which the relative density of a liquid will change per unit of temperature. For example, if a liquid expands by 1% by a one degree change in temperature, the expansion coefficient is one.

- *FEU:* Forty-foot equivalent unit, roughly the same as two 20-foot equivalent units though not exact.
- *Grain capacity:* The total amount of cargo space available for the carriage of bulk cargo. It includes the gaps between the frames and beams in the ship's hold.
- *Lane in metres:* A common measurement used on RORO vessels where vehicles are carried. Because of the different lengths of vehicles, it is easier to describe the space available as lane metres rather than to state the amount of medium- or small-sized cars that the vessel can carry.
- *Stowage factor:* The space occupied by 1 metric tonne of any given cargo. For example, the stowage factor of clinker is 1.5. This means that 1 tonne of clinker will occupy 1.5 m^3. It may also be expressed in cubic feet per tonne.
- *Tank volume:* The total volume of the tank minus an allowance made for the frames and beams. It is good practice not to fill the tank completely to allow for expansion of the cargo.
- *TEU:* Twenty-foot equivalent unit, the size of a typical shipping container.

2.2 TERMS USED FOR WEIGHT AND QUANTITY

Apart from cargo damage, there is no greater subject of dispute between carriers and shippers than that of cargo quantity. Even the slightest discrepancies can give rise to long and expensive legal battles between the contracting parties. To avoid this, several terms are used to denote specific weights or volumes. Various methods are used to calculate the amount of cargo loaded and discharged by a vessel. This often depends on the vessel's type, the type of cargo, and existing facilities both onboard and in port. The techniques for calculating the quantity or volume of cargo range from traditional tallying to draught surveys and electronic bar coding.

(1) *Tallying:* Tallying is normally carried out in ports where the risks of short landing or pilferage are prevalent. This task is usually carried out by the stevedoring company responsible for cargo operations and at times complemented by tallying by the ship's staff. The figures are compared with the amount recorded on the Bill of Lading. The Bill of Lading is then the official carriage document. Tallying is most performed in the general cargo trades.

(2) *Weighing:* Weighing of cargo is usually done onshore with vessels rarely having the ability to check the accuracy of the weights

provided. In general, cargo trades where the weights are provided by the shipper, the vessel can only estimate the accuracy of such figure by working out the volume of the cargo and its associated stowage factor. The only other method that can be used by the vessel is the draught survey, which we will discuss in detail later.

(3) *Ullaging*. Vessels carrying bulk liquid can cross-check shore figures by calculating the volume of cargo they have onboard, and thus the weight. Several types of gauges are available, with one of the most common being the float. Traditionally, as the height of the liquid rises, the ullage (i.e. the distance between the liquid surface and the tank ceiling) decreases. This height is measured. Tables are then consulted to yield the volume occupied by the cargo. The ambient temperature needs to be taken into consideration as any variation in temperature will affect the density of the cargo. Worked examples of that sort can be found at the end of this section. For seafarers employed on oil tankers, or when engaging in bunkering, it is worth noting that any increase in temperature will cause a decrease in the relative density (r.d.) of oil. In order that the weight may be calculated, it is important that when ullages are taken, the r.d. of the oil is also taken. The change of r.d. due to a change of one degree in temperature is known as the r.d. coefficient. Today, more modern methods may be available to measure the ullage. Whichever way the ullage is calculated, great care should be taken as the build-up of static electricity inside the tank can become a serious hazard for the vessel. To mitigate these hazards, sounding tapes are specially designed for ullaging petroleum tanks. Tankers normally have ullaging and sampling ports that reduce the risk of explosion. Electronic means of sounding the tanks may also be used. This is common on modern tankers and makes use of ultrasonic waves of energy to measure the level of the liquid inside the tank. Afterwards, tank calibration tables are used and corrections for temperature and density are applied to obtain the exact amount of cargo in each tank. The stowage factor of the cargo, that is, the volume occupied by the cargo, the broken stowage, and the relative density of the liquid or its expansion coefficient are some characteristics that the cargo officer must know to be able to carry out these calculations.

(4) *Using the calibration table:* Estimation of volume in the case of an unsymmetrical tank can be a complex matter. Ships' tanks are thus 'pre-measured' or calibrated when the ship is being built and the results tabulated for easy reference. Care should be taken when reading these tables as they may refer to either the ullage or the sounding. Sometimes corrections must be made depending on the location of

the ullaging or sampling pipe if the vessel is listed or on trim. Various other types of cargo measurement exist, most of which use the same principles as those described earlier. In any case, further reading is strongly encouraged.

2.3 DRAUGHT SURVEY

The objective of the draught survey is to first establish the weight of the vessel in its entirety, and then the non-cargo weights. The difference between the two will provide the weight of the cargo. This process is generally considered one of the most accurate forms of cargo measurement in the bulk trade. Draught surveying is an art that is backed up with science. It does require some practice and special attention to small details as these can affect the outcome quite considerably. The draught survey procedure is divided into three main steps:

(1) Reading of draughts and soundings
(2) Extracting information from the hydrostatic tables
(3) Assessing the underwater volume

The whole draught survey is based on the reading of the ship's draught. Therefore, no effort should be spared in ensuring that these are as accurate as possible. To ensure accuracy, the officer or surveyor must be as close to the draught marks as practicably possible. The draught must be measured to two decimal points, that is, to the nearest centimetre. Finally, the draught must be read, at a minimum, at the six draught marks. While the readings are being made, it is imperative that there is no transfer of liquid or movement of any weight, including the rigging of cranes or other lifting gear. In turbulent conditions, the wave pattern should be studied with a mean of the highest and lowest draught readings recorded. This should be done at least 12 times and repeated for each of the draught marks. Specialised draught measuring devices can be used to improve the accuracy of draught reading. These can even be constructed onboard using a piece of open-ended clear hose. In other instances, the use of a length of clear, rigid plastic tube stretching across the deck may be useful. The leeward side draughts are read, after which the windward side draughts are calculated using the improvised manometer. Modern vessels are typically fitted with draught gauges which operate electrically or by water pressure. It is important to note that, more often than not, draught gauges are unreliable. Therefore, never replace manual observation of the draught marks with draught gauges. Correcting the draught readings is necessary as the draught marks on the vessel do not normally correspond to the vessel's perpendiculars and because the data from the ship's hydrostatic tables are related to the true mean draught.

Confusion often appears to arise concerning readings obtained from the two main types of hydrometers used by the shipping industry. Hydrometers are used in the industry for two different purposes. The first is the load line hydrometer. To assess whether a ship is overloaded in terms of the *International Convention on Load Lines, 1966*, a government surveyor will use a load line hydrometer. This measures specific gravity, otherwise known as relative density (r.d.). Specific gravity is defined as the ratio of density of a sample (in our case seawater) at temperature T1 to that of pure water at temperature T2. This type of hydrometer, therefore, specifies two temperatures T1/T2, often at 15°C/15°C or 60°F/60°F. If the seawater temperature varies significantly from T1, then a small correction should be applied due to the fact that the hydrometer is not at its standard temperature and will therefore have experienced some expansion or contraction. The temperature correction is usually quite small and only needs to be applied in extreme cases. The other main purpose for the use of hydrometers is in draught survey work. The draught survey is carried out as part of the task to calculate the commercially accepted weight of cargo onboard a ship. The purpose-made draught survey hydrometer is calibrated to provide the commercially accepted weight in kilogrammes of 1 litre of seawater. This is weight in air and is often given the term 'apparent density'. These hydrometers are calibrated at a standard temperature, but for draught survey purposes, a temperature correction is not required. Whilst it is true that a small error is introduced due to the hydrometer not being at its standard temperature, this error is compensated for by a change in volume of the ship. This change is due to the same temperature difference but is opposite to slight 'apparent density' differences. Therefore, these two differences tend to cancel each other out. The rule, then, when carrying out a draught survey, is to make no temperature correction to the 'apparent density' reading.

As we mentioned earlier, the most commonly used load line hydrometers are those calibrated at 15°C/15°C or 60°F/60°F. The difference in calibration temperature does not affect the result obtained. The density of pure water at 15°C (60°F) is 0.9991 kilogrammes per litre. A seawater specific gravity reading of 1.025 on this instrument means that the seawater has an actual density of 1.025 × 0.9991 kilogrammes per litre or 1.0241 kg/litre. A 1-litre sample of seawater of this density will weigh 1.0241 kilogrammes in a vacuum. Commercial weights are those measured in air and 1 litre of seawater will, in air, have a buoyancy force acting on it. This force is 0.0011 kilogrammes for every litre. The commercial weight is therefore smaller than the weight in a vacuum by this amount. Therefore, the 1-litre sample will weigh 1.0241–0.0011 = 1.023 kilogrammes in air. The draught survey hydrometer is calibrated to give this reading. For the sample of seawater, the load line hydrometer shows a specific gravity of

1.025 and the draught survey hydrometer shows that the weight in air of 1 litre of seawater is 1.023 kilogrammes. Both are correct. Therefore, a reading taken on a draught survey is required to be altered by adding 0.002 to the reading to give the Load Line surveyor's specific gravity reading. The ship's officers, marine surveyors, draught surveyors, and other persons involved in the loading of the ship should be familiar with this difference so that no confusion arises in relation to overloading, stability calculations, or draught surveys.

The hydrometer is an essential piece of equipment that is used when conducting a draught survey. It measures the density of the liquid it is floating in. Without going into details of the equipment, attention is drawn to the fact that several types of hydrometers exist and, unless the correct one is used, errors can affect the final result. When using the hydrometer, it is important that good care is taken to ensure an accurate reading. This means the equipment should be clean and free of any oil residues on the surface. Always use a clean container to collect the sample to be measured. Protect the container and the equipment from direct sunlight and sudden temperature changes. Lower the hydrometer gently into the liquid, slightly pressing downwards and then releasing it. Avoid parallax when reading the instrument.

When sampling seawater density, there are a number of key factors that need to be considered to ensure the accuracy of the reading. Seawater samples should be taken at different levels and at different positions along the ship's hull. Remember that any mud in suspension can affect the density of water in some places. Try to avoid collecting samples in the vicinity of outlets and discharge valves and beware of stagnant water that is trapped between the ship and the quayside. Also, bear in mind that tidal variations may affect the water density. Remember, it is important to sample the density of the dock water immediately after reading the draughts. When the draughts have been read and a mean has been obtained for the forward, midship, and aft draught, the arithmetical mean draught must then be calculated. The following formula demonstrates how to do this. Note that a correction must be applied to the draught mark readings if these do not correspond to the forward and aft perpendiculars. These corrections are usually found in the ship's particulars. When obtaining the true mean draught, remember that a vessel trims about its Centre of Flotation (LCF). At this point, the vessel's draught does not change when the trim changes. For this reason, hydrostatic tables that are calibrated for True Mean Draught are measured at the LCF. To convert the Arithmetical Mean Draught to the True Mean Draught, a correction must be applied. This is known as the layer correction. The layer correction is found by using the first trim correction formula. This is as follows:

Layer correction = (trim × distance between midship and LCF)/LBP

Here:
Trim is measured in centimetres:
True Mean Draught – AMD ı Layer Correction

By entering the hydrostatic tables with the True Mean Draught, the following information can be found:

TPC   (tonnes per centimetre immersion)
MCTC (moment to change trim by 1 centimetre)
KB   (vertical distance between the Centre of Buoyancy (CB) and the keel)
LCB   (distance between the CB and aft perpendicular of midship)
LCF   (distance between the Centre of Flotation and aft perpendicular of midship)
KM   (vertical height of the initial metacentre above the keel) and the actual displacement can be found.

Having obtained the displacement and the relevant data corresponding to the True Mean Draught, some corrections need to be applied. These are necessary if the vessel is not on an even keel or is in dock water whose density is different from the one the tables are made for, or if the vessel is hogged or sagged. The first trim correction is a function of the layer correction. It can be additive or subtractive as has been explained earlier. However, the first trim correction must also be applied to the extracted displacement. The following formula is used to find the correction needed:

Displacement correction (tonne) = layer correction × TPC × 100

Once the first correction is calculated, it is necessary to work out the second trim correction. This is required as the centre of flotation is different from the one specified in the tables as the tables were made with the vessel on an even keel. This correction is always additive. It is important to note in this calculation that the trim is measured in metres. The second trim correction is also known as the Nemeto formula:

$$\text{Correction in tonnes} = \frac{50 \times \text{trim}^2 \times (MCTC2) - MCTC1}{\text{Length BP}}$$

Here:
MCTC 2 = the moment to change trim at a draught of TMD + 50 centimetres
MCTC 1 = the moment to change trim at a draught of TMD – 50 kilogramme.

2.3.1 Contextual Example

If a vessel whose LBP is 114 metres, and has a forward corrected draught of 6.34 metres, and an aft corrected draught of 8.92 metres, what would the second trim correction be according to the hydrostatic tables provided above?

Calculated trim = 2.58

$$\text{True Mean Draught} = \frac{(6.34 + 8.92)}{2} = 7.63\text{m}$$

MCTC (+50) = MCTC for a TMD of 8.13 metres = 320.02
MCTC (−50) = MCTC for a TMD of 7.13 metres = 265.20
MCTC 2−MCTC 1 = 320.02−265.20 = 54.82

$$\text{Correction} = \frac{(50 \times 2.58^2 \times 54.82)}{114}$$
$$= 160.05 \text{ tonnes}$$

The next correction is made if the vessel is floating in dock water whose density is different from that which the hydrostatic tables were made. Some tables are made for freshwater, others are made for seawater. If the density of the water is different, the density correction must be applied accordingly:

$$\text{Correction} = \frac{\text{Displacement} \times \text{Actual Density of Water}}{\text{Density of Sea (or Fresh Water)}}$$

Now that the displacement of the whole vessel has been calculated, the next step is to establish all non-cargo deductibles onboard. This consists mostly of fuel, freshwater, and ballast water. In this process, extreme care should be exercised when sounding the tanks, especially if the vessel is trimmed. In this case, a correction should be made to the sounding. Sounding corrections are usually published within the ship's particulars. The density of the liquid must be taken at the same time as the sounding. In some instances, the temperature of the liquid in the tank must be ascertained first. Subtracting the non-cargo deductibles from the corrected displacement will give the amount of cargo onboard.

One term which frequently causes confusion is the *constant*. This is the figure that remains when all non-cargo deductibles have been subtracted from the true displacement, that is, the difference between the calculated light ship displacement (from the ship's particulars) and the actual light ship displacement. It is called the 'constant' because it is assumed to be unchanged from the ballast condition at the commencement of loading until the end of loading. The figure can often change from voyage to voyage; this can be caused by changes in stores and the quantities of spares carried onboard, inaccuracies in determining fuel and ballast amounts and draught readings, as well as by changes to the ship's structure (for example, the accumulation of paint over time, additions and removals of metal work, and so forth). Refer to the following example of the *MV Kagoro*. In this example, the light displacement is 22,704 gross tonnes. The tabulated displacement is 22,250 gross tonnes. By subtracting the light displacement from the tabulated displacement, we can see the tabulated light displacement is 454 gross tonnes. The *constant*, therefore, is 454 gross tonnes.

In summary, to carry out an accurate draught survey, it is important to read the ship's draught at each of the draught marks. Remember to obtain dock water samples. Establish the density of the dock water and average out if necessary. Average the draught readings for the forward, midship, and aft. Correct the draught readings for readings at the perpendicular. Find the arithmetical mean draught. Ascertain whether the vessel is hogging or sagging. Establish the true mean draught by applying layer corrections to the AMD. Using the hydrostatic tables, extract the correct data including the displacement. Apply first trim corrections to the extracted displacement, followed by second trim corrections to the extracted displacement. Apply the density correction to the extracted displacement. Calculate the corrected displacement. Sound all the tanks and work out the volume carried. Find out the density and temperature of liquids and calculate the weights contained in each tank. Finally, subtract the deductibles from the corrected displacement. The result is the amount of cargo onboard. It is worth mentioning that there are other methods used to conduct draught surveys, all of which yield very slight differences from the method outlined here.

This then concludes this chapter on determining cargo quantity and volume. In Chapter 3, we will begin to examine ballast management, which we briefly touched on in Chapter 1. Before starting the next chapter, it is probably worth spending a few minutes acquainting yourself with the worked example provided in Figures 2.1 and 2.2. Once you are confident with the figures given, try changing the data and calculating your own displacements.

U.N.   • ECE • DRAUGHT SURVEY CODE

DRAUGHT SURVEY REPORT OF CARGO IN BULK	☐ LOADED ☐ UNLOADED	FORM 'C'	

Corporate
Identification:
Vessel M/V: Survey No:

DRAUGHT STATEMENT

DRAUGHT READINGS HOURS:	STARTING SURVEY FROM: TO:	FINISHING SURVEY FROM: TO:	
	Metres	Metres	
Draught forward port	4.92	10.08	From observation
Draught forward starboard	4.93	10.08	From observation
Draught forward mean	4.925	10.080	(Line 1 + line 2) / 2
			From ship's particulars
Stem correction	- 0.057527	-0.00236	Line 3 + line 4
Draught forward (corrected to fore pp.)	4.867473	10.077639	
Draught after port	7.24	10.17	From observation
Draught after starboard	7.24	10.18	From observation
Draught after mean	7.240	10.175	(Line 6 + line 7) / 2
Stem correction	0.089100	0.00366	From ship's particulars
Draught forward (corrected to fore pp.)	7.329100	10.178660	Line 8 + line 9
Draught fore and after mean	6.098287	10.128150	(Line 5 + line 10) / 2
Draught midship port	5.88	10.10	From observation
Draught midship starboard	5.90	10.12	From observation
Draught midship mean	5.890	10.110	(Line 12 + line 13) / 2
Midship correction	0.013378	0.000549	From ship's particulars
Draught midship (corrected to midship pp)	5.903378	10.110549	Line 14 + line 15
Sag (+) Hog (-)	- 0.194908	-0.01760	Line 16 - line 11
Mean of means	6.000832 3	10.119349	(Line 11 +Line 16) / 2
Draught extreme corrected for hog/sag	5.952105 1	10.114949	(Line 16x6 + Line 5 + line 10) /8
Correction (-) for keel thickness if applicable	0	0	From ship's particulars
Draught moulded corrected for hog/sag	5.952105 1	10.114949	Line 19 - line 20
Trim: fwd (-) aft (+)	246.2	10.01	Line 10 - line 5. Note: this figure in cm
	Kg/m³	Kg/m³	
Observed density	1012	1009	From observation
Ship's tables density Kg/m3			
Hydrometer No.			
	Metric tonnes	Metric tonnes	
Displacement at Kg/m³	18,306.0 0	32,162.00	
First trim correction	130.27	2.12	Trim (cm) x TPC X LCF/LBP; from ship's particulars
Second trim correction	18.61	0.02	Trim (m) 2 x 50 x MTC diff/LBP; from ship's particulars
Total trim correction	148.88	2.14	Line 25 + line 26
Displacement corrected for trim	18,454.8 8	32,164.14	Line 24 + line 27
Correction for density average	-234.06	-502.07	(1025 - line 23) / 1025 x line 28 (negative if density less than 1025)
Displacement corrected for density	18,220.8 2	31,662.07	Line 28 + line 29
Total deductibles	9,381.20	989.00	From ship's figures
Displacement corrected for deductibles	8,389.62	30,673.07	Line 30 - line 31
TOTAL CARGO		tonnes	Line 32 finishing - line 32 starting

Draughts, densities, fresh water and ballast soundings witnessed and agreed to by the Chief Officer. Fuel oil soundings witnessed and agreed to by the Chief Engineer unless otherwise stated in form 'A'.

Figure 2.1 Worked example of a draught survey.

Source: Author's own.

CARGO SUMMARY – LOADING

SHIP	:	MV KAGORO	BERTH	:	KOORAGANG 4
CARGO	:	KOORAGANG COALS	Commencing 2324 hours on 8 Jan 22		
PORT	:	NEWCASTLE, AUS	Completed 1114 hours on 10 Jan 2022		
		For a JAPANESE PORT			

MASTER :

CHIEF OFFICER :

DENSITIES		Commencing		Completion	
Observed Density		1019.0		1022.5	
Tempr Correction		.0		.0	
Corrected Density		1019		1022.5	

DRAUGHTS	Port	Starboard	Port	Starboard
Forward	6.70	6.70	15.07	15.08
Aft	9.86	9.86	15.27	15.28
Mean		8.280		15.175
Midship	8.04	8.08	15.21	15.26
Mean of Midship		8.060		15.235
Mean of Means		8.170		15.205
Mean of Means of Means		8.115		15.220

	Commencing	Completion
Displacement (Tabulated)	81986	162128
Trim Correction	-1550	-18
Subtotal	80436	162110
Density Correction	-471	-395
Displacement (Corrected)	79965	161715

Figure 2.2 Cargo loading and discharge summary.

SHIP	:	MV KAGORO	BERTH	:	KOORAGANG 4
CARGO	:	KOORAGANG COALS			Commencing 2324 hours on 8 Jan 22
PORT	:	NEWCASTLE, AUS			Completed 1114 hours on 10 Jan 2022
		For a JAPANESE PORT			

MASTER :

CHIEF OFFICER :

DISPLACEMENT (TONNES)
====================

	COMMENCING	COMPLETION
	79965	161715
Fuel Oil	1025	1015
Diesel Oil	101	94
Freshwater (Domestic)	328	306
Freshwater (Cargo)		
Ballast	55807	51
Cargo Other		
	--------	--------
Total Variables	57261	1466

DISPLACEMENTS:	Corrected Light	22704	Corrected Loaded	160249
	Tabulated Light	22250	Corrected Light	-22704
		--------		--------
		454	CARGO LOADED	137545
Vessel Constant 'K':		tonnes		

cms

REMARKS
========

VSL LOADED FOR 1.60m TIDE – ALL HOLDS FILLED TO CAPACITY EXCEPT #7 AS
TIDAL DRAFT OF 15.29m HAD BEEN REACHED AFT
SHORE TONNAGE – 137412 tonnes
SHIP TONNAGE – 137545 tonnes
MAX S/F 35% AT FR. No. 319
MAX B/M 54% AT FR. No. 305
STRESSES WELL WITHIN ALLOWABLE THROUGHOUT LOADING

SHIP   :   MV KAGORO
CARGO   :   KOORAGANG COALS
PORT   :   NEWCASTLE, AUS

BERTH   :   KOORAGANG 4

For a JAPANESE PORT

Commencing 2324 hours on 8 Jan 22

Completed 1114 hours on 10 Jan 2022

MASTER   :

CHIEF OFFICER :

LOADING SEQUENCE & QUANTITIES
==============================

HATCH	DESCRIPTION	R/TONNES	R/TONNES	R/TONNES	RUN TOTAL
1.	BAYSWATER	3/8000	11/5000		23000
2.	BIG BEN	8/16400			16400
3.	HOWICK	6/17300			17300
4.	HUNTER VALLEY	2/10141	13/6200		16341
5.	BAYSWATER	4/10000	10/5700		15700
6.	BIG BEN	9/16300			16300
7.	HUNTER VALLEY	1/10000	12/5000	14/400	15400
8.	PEKO	5/15571			15571
9.	HUNTER VALLEY	7/11400			11400
10.					
11.					
12.					

TOTALS   SHIP   137545   SHORE 137412

SHIP/SHORE = 100 %

Figure 2.2 (Continued)

Chapter 3

Ballast Management

3.1 INTRODUCTION

In the previous chapters, we have touched very lightly on ballasting already. We know that ballasting is a critical action that is taken to maintain the vessel's trim during cargo loading and discharging. In this chapter, we will examine ballasting a bit further, and more importantly, how ships manage their ballast in relation to their cargo operations. Each year, over 120 million tonnes of ballast water are discharged within UK waters. The carriage of ballast onboard provides the vessel with adequate stability, especially when the vessel is only part loaded. Ballasting and de-ballasting are important aspects of cargo operations, and especially so on bulk carriers, container ships, and tankers. On bulk carriers, good ballast management reduces the risk of causing excessive stress to the structure of the vessel. On tankers, ballast water is normally carried in dedicated tanks, which reduces the risk of contamination that would otherwise occur if the ballast water was pumped directly into the cargo tanks. This is important as under Annex I of the *International Convention for the Prevention of Pollution by Ships*, or the MARPOL Convention, which was adopted by the IMO in 1973, and subsequently amended in 1978, ships are prohibited from discharging oil or oily water overboard. Apart from architectural reasons, this is one of the main reasons why ships have separate ballast tanks.

3.2 RISKS INVOLVED WITH THE EXCHANGE OF BALLAST WATER

The carriage of ballast water, however, is not without its inconveniences. Harmful microorganisms live in seawater and when this water is pumped into the ship's ballast tanks, the microorganisms are sucked in as well. Various types of exotic species, including fish, molluscs, worms, and toxic algae, have been found in British waters, more often than not the result of ballast water discharges. The problem is so widespread that the introduction

DOI: 10.1201/9781003354338-4

of non-indigenous marine species is now recognised as a significant world-wide problem. One typical example is dinoflagellates, which are a species of toxic algae. Dinoflagellates are extremely harmful to the marine environment and have been known to cause serious diseases and spread parasites which affect humans, animals, and plants. By upsetting the natural ecological order, dinoflagellates impact on native species through cross-breeding and habitat destruction. This in turn can increase pollution through blocked waste pipes and the foul odours. As we said, dinoflagellates are sucked into the ship's tanks when ballast water is pumped in and remain alive in the vessel tanks until the ballast water is pumped out (usually during loading in port). These organisms then enter the feeding cycle producing toxins, which can cause paralysis or even death in human beings when infected shellfish are eaten. In an attempt to preventing the transfer of harmful aquatic organisms, on 13 February 2004, the IMO adopted the *International Convention for the Control and Management of Ships' Ballast Water and Sediments* (BWM), with the convention coming into effect on 8 September 2017. The exchange of ballast water at sea has since been recognised as a means of preventing the introduction of harmful aquatic organisms. There are two recognised methods that ships can undertake when performing ballast water exchange at sea: (1) the flow-through method and (2) the exchange of ballast in deep waters.

The flow-through method, which was developed following research carried out on the *Iron Whyalla*, has been adopted as one of the safest methods for exchanging ballast. The process consists of flushing the ballast tank out with three times its water capacity. This action removes up to 95% of the original ballast water, thus reducing the risk of environmental pollution. However, for the process to be very efficient, some tanks need to be modified to allow for maximum water mixing. The flow-through method remains the most commonly used method on ships today. It is recommended that water be taken from areas where the water depth exceeds 2,000 metres (6,561 ft) as at that depth, it is assumed the water will contain fewer organisms, which are able to survive in freshwater and coastal water. The second method is the more traditional process of emptying the ballast tanks and refilling them with ballast water obtained in deep waters. This method has its limitations since the emptying of any tank at sea creates free surface moments. Free surface moments are dangerous as they reduce the vessel's stability. This method may also create larger shearing forces and bending moments on the vessel's structure. If these forces are greater than the metal can tolerate, seams can crack eventually leading to leaks and distortional damage to the ship's structure.

3.3 BALLAST WATER MANAGEMENT GUIDELINES

The *International Convention for the Control and Management of Ships' Ballast Water and Sediments*, or *Ballast Water Management Convention*

(BWM) for short, entered effect on 8 September 2017. The Convention applies to all vessels that operate in waters of more than one signee to the Convention. In other words, all vessels engaged in international trade must comply with the BWM Convention. The Convention applies to all vessels, regardless of size or tonnage, that are entitled to fly the Flag of a signee to the Convention. There are however a few significant exceptions, whereby ships do not have to comply with the BWM Convention:

(1) Ships not constructed and/or designed to carry ballast water.
(2) Ships that only operate in the waters of a single signee to the convention (i.e. domestically operating vessels).
(3) Ships operating in the waters of a single signee and on the high seas.
(4) Warships, naval auxiliary, or ships owned or operated by a State and used only on government non-commercial service.
(5) Permanent ballast water in sealed tanks on ships that are not subject to discharge.

The Convention defines a 'ship' as a vessel of any type operating in the aquatic environment and includes submersibles, floating craft, floating platforms, floating storage units (FSUs), and floating production, storage, and offloading vessels (FPSOs). Floating craft encompasses a wide variety of vessels that operate in the marine environment and use ballast water for stability, heeling, or operating purposes and includes, but is not limited to, fishing vessels, large yachts, dumb barges, and so forth. Ships subject to the Convention requirements are obliged to conduct ballast water management in accordance with the provisions stated within the BWM Convention. These provisions are outlined in the following sections.

3.3.1 Ballast Water Management Plan

Ships must carry and implement a Ballast Water Management Plan (BWMP) that has been approved by the vessel's Administration. For example, in the UK, this is the UK Maritime and Coastguard Agency (UK MCA), or in Australia, the Australian Maritime Safety Authority (AMSA). The plan must include details of the safety procedures for the ship and crew and provide a detailed description of the actions to be taken to implement the ballast water management requirements. Further information is provided in IMO Guideline G4.

3.3.2 Ballast Water Record Books

Ships must carry a Ballast Water Record Book (BWRB), which must be completed following each ballast water operation. The form of the Ballast Water Record Book should emulate that contained in Appendix II of the BWM

Convention. In the UK, Ballast Water Record Books are available for purchase from the HM Stationery Office.

3.3.3 Ballast Water Management Standards

The BWM Convention, as amended, introduces the phased implementation of two Ballast Water Standards:

> D1 – Ballast Water Exchange Standard
> D2 – Ballast Water Performance Standard

Currently, any ballast water discharged from a ship should meet either the D1 or D2 standard until such time as the vessel is required to implement the D2 standard. Ships currently meeting the D2 standard (usually through the use of a ballast water treatment system) can opt to meet the D1 standard, though it is recommended that any ballast water treatment equipment installed onboard is used to comply with the D2 standard. The BWM Convention implementation schedule means that the use of ballast water exchange, which meets the D1 standard, as a management method, will be replaced by a requirement for ballast water to meet the D2 discharge performance standard (usually through the use of a ballast water treatment system).

3.3.4 Sediment Management for Ships

All ships must remove and dispose of sediments from spaces designed to carry ballast water in accordance with the ship's Ballast Water Management Plan.

3.3.5 Duties of Officers and Crew

The ship's officers and crew should be fully familiar with their duties with respect to the implementation of the ship's Ballast Water Management Plan and be competent to execute the plan accordingly.

3.3.6 Exceptions

The requirements to meet the ballast water management standards do not generally apply to the following:

(1) The uptake and discharge of ballast water necessary for ensuring the safety of the ship in emergency situations.
(2) The accidental discharge or ingress of ballast water because of damage to the ship or its equipment.

(3) The uptake or discharge of ballast water for the purpose of avoiding or minimising pollution incidents from the ship.

(4) The uptake and subsequent discharge on the high seas of the same ballast water.

(5) The discharge of ballast water from a ship at the same location where the whole of the ballast water originated, provided no mixing of unmanaged ballast water from other areas has occurred. If mixing occurs, the ballast water is subject to management in accordance with the convention.

3.3.7 Exemptions

Exemptions to the requirement to meet the management standards may be granted in specific circumstances. Exemptions may only be granted to a ship or ships on voyage(s) between specified locations, ships which operate within a defined area, or to a ship that operates exclusively between specified locations. An exemption is usually effective for no longer than five years. The exemption can only be granted if ballast water is not mixed, other than in the locations specified on the exemption, and must be based on a detailed risk assessment which takes into account the relevant sections of the IMO Guidelines (i.e. Guideline G7).

3.3.8 Equivalent Compliance

Vessels used solely for recreation, competitions, or watercraft used primarily for search and rescue operations, that are less than 50 metres (164 ft) in overall length, and have a maximum ballast capacity of 8 m^3 may apply to their Administration for *equivalent compliance*. The decision to grant equivalent compliance is determined based on the relevant guidance developed by the IMO – that is, the Guidelines for Ballast Water Management Equivalent Compliance (G3). There is no other equivalent compliance currently available under the BWM Convention.

3.4 SURVEY AND CERTIFICATION OF SHIPS

Ships of 400 gross tonnes and above are subject to the survey and certification regime as stipulated within the BWM Convention. Vessels under this threshold are still required to meet the requirements of the BWM Convention, and as such Administrations are required to establish appropriate measures to ensure compliance by vessels of less than 400 gross tonnes.

3.5 BALLAST WATER MANAGEMENT STANDARDS

The BWM Convention requires that ballast water is managed to meet the standards set by the BWM Convention and allows for the phased introduction of two standards as detailed under Regulations D1 and D2. D1 details requirements relating to ballast water exchange and D2 details allowable limits for organisms within the ballast water discharge. The BWM Convention allows for D1 to be used until such time as D2 is required but does not prevent ships operating to the D2 standard ahead of schedule for implementation.

3.5.1 D1 – Ballast Water Exchange (BWE)

The standard set by the BWM Convention states that ships undertaking ballast water exchange must do so with an efficiency of at least 95% volumetric exchange of ballast water. For ships exchanging ballast water by the pumping-through method, pumping through three times the volume of each ballast tank is generally considered equivalent to meeting the 95% standard. Ships undertaking ballast water exchange should conduct the operation at least 200 nautical miles (230.1 mi, 370.4 km) from the nearest land and in water at least 200 metres (656 ft) deep; or in cases where the ship is unable to conduct ballast water exchange in accordance with the above, as far from the nearest land as possible, and in all cases at least 50 nautical miles (57.3 mi, 92.6 km) from the nearest land and in water at least 200 metres (656 ft) deep. In sea areas where the minimum distance and depth criteria cannot be met, the signees to the Convention have the ability, within their waters, to designate ballast water exchange–specific areas. Areas designated by a signee should be used in compliance with the terms of use stipulated by the Administration(s) responsible for the designation. Vessels may be required to deviate or delay their voyage to use the designated ballast water exchange area. Shipowners and vessel operators are urged to contact the relevant Port State Authority for confirmation of ballast water exchange requirements within local waters.

3.5.2 D2 – Ballast Water Performance Standard

D2 stipulates the acceptable level of organisms that may be found within discharged ballast water. The D2 Standard specifies that treated and discharged ballast water must have fewer than ten viable organisms greater than or equal to 50 micrometres in minimum dimension per cubic metre, and fewer than ten viable organisms less than 50 micrometres in minimum dimension and greater than or equal to 10 micrometres in minimum dimension per millilitre. In addition, a ballast water discharge of indicator microbes, as a health standard, should not exceed the following specified concentrations:

   (1) Toxicogenic *Vibrio cholerae* (O1 and O139) with less than one colony-forming unit (cfu) per 100 millilitres or less than 1 cfu per 1 gram (wet weight) zooplankton samples.

(2) *Escherichia coli* less than 250 cfu per 100 millilitres.

(3) Intestinal Enterococci less than 100 cfu per 100 millilitres.

Ballast water treatment equipment is developed, and type approved on the basis of the equipment's ability to treat the ballast water to the required standard. Although not the only way to meet the D2 standard, the installation of an appropriately type approved ballast water treatment system is the most common method used to comply with the Convention.

3.6 OTHER METHODS OF BALLAST WATER MANAGEMENT

Other methods may be accepted as alternatives to either D1 or D2 provided the methods ensure at least the same level of protection to the environment, human health, property, or resources and are approved in principle by the IMO.

3.7 BALLAST WATER MANAGEMENT IMPLEMENTATION SCHEDULE

The requirement to meet either D1 or D2 standards does not apply to ships that discharge ballast water to a reception facility that has been designed taking into consideration Guideline G5: Guidelines for Ballast Water Reception Facilities. Ships are required to meet either the D1 or D2 standard until such time as they are required to meet D2. Table 3.1 outlines the implementation dates for the D2 standard.

3.8 UK IMPLEMENTATION OF THE BALLAST WATER MANAGEMENT CONVENTION

Further information regarding the UK implementation of the Ballast Water Management Convention can be found at www.gov.uk/government/publications/ballast-water-management-faq

3.9 BALLAST WATER MANAGEMENT REQUIREMENTS IN SPECIFIC MARINE AREAS

In addition to the BWM Convention, specific areas of the world's oceans have additional requirements for ballast water control, which vessels must adhere to. These include in waters around the United States (US) and the Antarctic:

Table 3.1 Ballast Water Management Implementation Schedule

D1 or D2 compliance	EIF – Entry into Force of the Ballast Water Management Convention
D2 compliance	IOPP Renewal survey of the International Oil Pollution Prevention Certificate

Entered into force (EIF) on 8 September 2017

	8/9/17–7/9/18	8/9/18–7/9/19	8/9/19–7/9/20	8/9/20–7/9/21	8/9/21–7/9/22	8/9/22–7/9/23	8/9/23–7/9/24	8/9/24 Onwards
	1st IOPP renewal following EIF					D2 compliance (2nd IOPP renewal following EIF)		
		1st IOPP renewal following EIF					D2 compliance (2nd IOPP renewal following EIF)	
			D2 compliance (1st IOPP renewal following EIF)					
					D2 compliance (1st IOPP renewal following EIF)			
						D2 compliance (1st IOPP renewal following EIF)		
								D2 compliance

Vessels that do not hold an IOPP Certificate
Implementation schedule to be determined by administration but no later than 8/9/24

Ships constructed on or after EIF – D2 compliance

- *United States:* Ships trading in US waters should be aware that ballast water management requirements within US waters differ from those outlined in the Ballast Water Management Convention. Ships staff are advised to contact the relevant authority for further information.
- *Arctic and Antarctic waters:* There are specific requirements for the uptake or discharge of ballast water in Arctic or Antarctic waters. You must follow these unless the safety of the ship is jeopardised by a ballast exchange, or where it is necessary for saving life at sea. A ballast water management plan must be prepared for vessels entering Antarctic waters, taking into account problems of ballast water exchange in Antarctic conditions. The vessel should keep a record of ballast water operations onboard. Ballast water should first be exchanged before arrival in Antarctic waters or at least 50 nautical miles (57.5 mi, 92.6 km) from the nearest land and in waters that are at least 200 metres (656 ft) deep. Similarly, ballast water taken on in Antarctic waters should be exchanged north of the Antarctic Polar Frontal Zone, and at least 200 nautical miles (230.1 mi, 370.4 km) from the nearest land in water that is at least 200 metres (656 ft) deep. The release of sediments during the cleaning of ballast tanks should not take place in Antarctic waters. Vessels that have spent significant time in the Arctic should discharge and clean tanks before entering Antarctic waters. If this is not possible, sediment accumulation in ballast tanks should be monitored and the sediment disposed of in accordance with the ship's Ballast Water Management Plan.

3.10 CASE STUDY: AUSTRALIAN BALLAST WATER MANAGEMENT REQUIREMENTS

The reason for the introduction of the mandatory Australian ballast water management arrangements is to help minimise the risk of the introduction of harmful aquatic organisms into Australia's marine environment through ship's ballast water.

3.10.1 Background

The Australian Quarantine and Inspection Service (AQIS) is the lead agency for the management of international vessels ballast water. Australia was the first country in the world to introduce voluntary ballast water management guidelines for international shipping, which have been in use by since 1991. In September 1999, the Australian Government announced that mandatory ballast water management arrangements would be introduced for all international vessels arriving in Australian ports or waters from 1 July 2001. The new arrangements incorporated a Decision Support System (DSS), which provides vessels with a risk assessment of the ballast water carried onboard as to the likelihood of introducing exotic species into Australian ports or waters. A revised ballast water reporting system and verification inspections was also implemented as an integral part of the new arrangements. The mandatory Australian ballast water management requirements were developed to be consistent with the IMO Guidelines for minimising the uptake of harmful aquatic species when vessels are performing ballasting operations. Australia's ballast water management requirements have legislative backing and are legally enforced under the *Quarantine Act of 1908*. Safety of vessels and crew are of paramount importance; therefore, vessels undertaking ballasting operations to meet Australia's ballast water management requirements must do so in accordance with the IMO Guidelines.

3.10.2 What the New Arrangements Mean for the Shipping Industry Mandatory Ballast Water Management Requirements

From 1 July 2001, all international vessels are required to manage their ballast water in accordance with AQIS requirements and not discharge high risk ballast water in Australian ports or waters.

3.10.3 Ballast Water Management Options

The ballast water management options approved by AQIS that vessel masters may undertake to minimise the risk of introduction of harmful aquatic organisms into Australian ports or waters are as follows:

(1) Non-discharge of 'high risk' ballast tanks in Australian ports or waters. This method may be employed where the vessel does not need to discharge any ballast water in Australian ports or waters, or where the vessel has undertaken a DSS risk assessment and the risk assessment was 'low'.

(2) Tank-to-tank transfer: This method may be employed where the vessel is able to move high-risk ballast water from tank to tank within the vessel to avoid discharging high-risk ballast water in Australian ports or waters.

(3) Full ballast water exchange at sea using one of the following methods:

- Flow-through method.
- Sequential method (empty/refill).
- Dilution method.

Full ballast water exchange may be employed where the vessel has high-risk ballast water intended for discharge in Australian ports or waters. Vessels should conduct full ballast water exchange in deep mid-ocean water, as far as possible from shore, and outside the Australian 12 nautical mile (13.8 mi; 22.22 km) limit. Exchange at sea must be undertaken to a minimum 95% volumetric exchange and should be undertaken in water greater than 200 metres (656 ft) in depth. Where full ballast water exchange cannot be undertaken due to safety reasons, such as weather, sea conditions, or operational impracticability, the master must report this to AQIS on the Quarantine Pre-Arrival Report (QPAR) as soon as possible and prior to entering Australian waters. Other comparable treatment methods may be considered by AQIS on a case-by-case basis. Vessel masters must contact AQIS prior to undertaking any treatment methods other than those specified above.

3.10.4 The AQIS Decision Support System

The Australian Ballast Water Decision Support System (DSS) is a computer software application developed by AQIS in consultation with the Australian maritime industry. The DSS undertakes a biological risk assessment that predicts the likelihood of entry of harmful aquatic organisms and pathogens on a tank-by-tank basis based on uptake and discharge information entered by the vessels master or agent. Information can be lodged with the DSS at the last port of call or as early as possible prior to entering Australian waters (12 nautical mile (13.8 mi; 22.22 km) limit). After submitting information into the DSS, the vessel receives a risk assessment number (RAN) which must be entered on the vessel's QPAR. This allows AQIS officers to search the DSS for the risk assessment when undertaking a verification inspection of the vessel. Vessel masters are encouraged to use the DSS for 'scenario testing' to allow the best possible ballast water management option for the vessel. Low-risk ballast water typically requires minimal treatment prior to discharging in Australian ports or waters. Entering information as early as possible into the DSS will allow masters more time to perform an AQIS-approved treatment prior to arrival in Australia. This saves time, money, and inconvenience. Access to the DSS is through either of the following methods:

(1) The online AQIS portal.
(2) Inmarsat-C or email.

3.10.5 Ballast Water Reporting

All vessels arriving in Australia from international waters are required to submit a QPAR to AQIS. The QPAR details the condition of the vessel, including human health, cargoes, and ballast water management. Vessel masters or agents are required to send the QPAR to AQIS between 12 and 48 hours prior to arrival in Australia. This allows efficient processing of the QPAR and avoids any disruption to the vessels arrival. Vessel masters or agents that do not submit the QPAR to AQIS are not given formal quarantine clearance to enter port. This has the unwanted effect of causing considerable delays to the vessel and inevitably incurs additional AQIS charges. Vessels require written permission to discharge any ballast water in Australian ports or waters which may be given the following lodgement of the QPAR with AQIS. If the vessel's ballast water details change, a revised

QPAR must be submitted to AQIS prior to the discharge of any ballast water. Vessel masters are required to complete two additional AQIS forms:

(1) *The AQIS Ballast Water Uptake/Discharge Log:* This log can also be used to provide the shipping agent with uptake and discharge information for entry into the DSS.
(2) *The AQIS Ballast Water Treatment/Exchange Log:* This log must be used to record all ballast water treatment/exchanges at sea.

These forms should not be sent to AQIS, although they must be held on the vessel for a period of two years and produced to AQIS on request.

3.10.6 Verification Inspections

AQIS Officers typically conduct ballast water verification inspections onboard vessels to ensure compliance with Australia's ballast water management requirements. AQIS Officers use the QPAR/DSS results, the AQIS Ballast Water Logs, and the vessel's deck and engineering logs to verify the information supplied to AQIS is correct. The verification inspection usually takes about 30 minutes to complete, and in most cases it is conducted at the same time as routine vessel inspections. Vessels that have a poor quarantine history or have not previously complied with AQIS requirements are almost always subject to AQIS inspections on each visit to Australia.

3.10.7 AQIS Ballast Water Compliance Agreements

AQIS Ballast Water Compliance Agreements are available to vessels who regularly visit Australian ports and who have demonstrated good quarantine compliance history. The Agreement sets out the details of the activities, how they are conducted, and who has responsibility for ensuring they comply with AQIS requirements. Ballast Water Compliance Agreements are subject to formal audit by AQIS on a regular basis.

3.10.8 Tank Stripping

The discharge of ballast tank sediment is prohibited in Australian waters. Ballast tank stripping must not occur where this operation involves the discharge of sediment in Australian waters. Written approval from AQIS must be obtained prior to performing ballast tank stripping or sediment removal.

3.10.9 Access to Sampling Points

All vessels visiting Australian ports of call are required to provide access to safe ballast water sampling points within the vessel. Ballast water samples may be required to ensure compliance with Australia's ballast water management requirements or for further ballast water research (Figure 3.1).

BALLAST WATER UPTAKE / DISCHARGE LOG

Commonwealth of Australia *Quarantine Act 1908* Section 27A

Vessel Name: _____ IMO / Lloyds No.: _____ Call Sign: _____

Master's Signature: _____ Date: _____ PAGE: ____ of ____

Ballast Water Tanks or Cargo Holds	BALLAST WATER SOURCE				BALLAST WATER SEA SUCTION STRAINERS				BALLAST WATER DISCHARGE		
	Each tank will require the ballast water uptake port to be recorded below	If the ballast water uptake was not conducted in a port, it must be recorded as latitude and longitude in the table below	Record the dates and times of ballast water uptake below		Provide the date when the BW sea suction strainers were last inspected ___/___/___ Are the BW sea suction strainers used in the ballasting operation in good order and repair? Y☐ N☐	Was a sea suction strainer used during uptake? circle Y or N (optional) 0 (e.g. S1 or PS1)		If ports of call change or information details for the voyage have altered, please submit an amended form to your agent for a further AQIS DSS risk assessment – mark it 'AMENDED' and include the revision date (*Quarantine Act 1908* Section 29)		
	Name of BW uptake Port	Give latitude and longitude in (degrees and minutes)	Date(s) BW was taken on board	BW uptake start times using local time (24hour clock) (optional)	BW discharge - circle partial or full discharge	YES or NO	Strainer ID	Australian port(s) of discharge	Date(s) of BW discharge	Estimated time BW discharge will finish (optional)
1		Lat / Long	/ /	S :	Partial / Full	YES NO			/ /	F :
2		Lat / Long	/ /	S :	Partial / Full	YES NO			/ /	F :
3		Lat / Long	/ /	S :	Partial / Full	YES NO			/ /	F :
4		Lat / Long	/ /	S :	Partial / Full	YES NO			/ /	F :
5		Lat / Long	/ /	S :	Partial / Full	YES NO			/ /	F :
6		Lat / Long	/ /	S :	Partial / Full	YES NO			/ /	F :
7		Lat / Long	/ /	S :	Partial / Full	YES NO			/ /	F :
8		Lat / Long	/ /	S :	Partial / Full	YES NO			/ /	F :

List multiple tanks / sources separately

CONTAINER SHIPS SEE SPECIAL INSTRUCTIONS

BALLAST WATER TANK CODES: Forepeak = FPT Aft peak = APT Double bottom = DB Bottom side tank = BT Bottom side tank = BST Deep tank = DT Wing tank = WT Top side tank = TST Cargo hold = CH Heeling tank = HT Water ballast tank = WBT Port = P Starboard = S Centre = C Bilge = BGT Other = O (specify)

Form 026A - Date of Effect 1 July 2001

Ships completing this AQIS BW Uptake/Discharge log must also enter the ballast water information into the ship's deck and engineering logbooks. A ship's logbook must be made available for inspection by a Quarantine Officer at any Australian port or any location within the Australian 12nm limit.

Figure 3.1 Model ballast water uptake/discharge log.

Source: Author's own.

Chapter 4

Principles for the Safe Handling, Stowage, and Carriage of Dry Cargoes

4.1 INTRODUCTION

As we mentioned earlier, there are many different types of cargo that are transported around the world such as general cargoes, unitised or containerised cargo, bulk cargo in liquid or solid form, refrigerated cargoes, and RORO cargo. Due to their nature, these require varying degrees of attention. Underpinning the basic principles of handling, stowing, and carrying cargo are four main factors, which must be considered:

(1) Full use should be made of the vessel's carrying capacity with broken stowage kept to an absolute minimum.
(2) Damage to the vessel must be prevented at all reasonable costs; this includes ensuring adequate stability and reducing structural stresses.
(3) Damage to cargo must be prevented or reduced to the absolute minimum.
(4) Proper segregation and equal distribution of cargo within the ship's holds must be maintained to reduce the risks of cargo damage and facilitate efficient cargo discharge.

In this chapter, we will discuss the loading, carriage, and discharge of general, unitised, and containerised cargoes.

4.2 WHAT IS GENERAL CARGO?

The term 'general cargo' comprises any cargo or goods which are transported in bags, bales, cases, crates, drums, bundles, or large pieces of machinery and vehicles. Reels of paper and Intermediate Bulk Containers of goods also typically fall within this category. As general cargo comes in different forms and shapes, it calls for special attention when being stowed in the ship's cargo holds. Because of the irregular shapes of some packages (for example, pieces of machinery, barrels, and steel coils), broken stowage

DOI: 10.1201/9781003354338-5

can be enormous. Furthermore, if the shape of the compartment is in itself irregular, this can aggravate losses for the shipowner. Subsequently, extreme care should be exercised at the planning stage. This might include going to the warehouse, checking the shape and measurements of the packages, and comparing them with the space available onboard. It should be noted this problem does not usually arise with cargo such as bagged cargo or pallets of cartons being stowed in regular-shaped compartments. General cargo ships were mostly prevalent prior to the advent of containerised shipping, and has since the 1960s, has fallen in terms of overall tonnage. That being said, general cargo ships are still active in parts of Asia, West and East Africa, and throughout South America.

4.3 EQUIPMENT USED IN THE HANDLING OF GENERAL CARGO

In conjunction with the second factor mentioned, that is, damage to the vessel, this can be partly mitigated by using the correct equipment to handle and stow general cargoes. Some examples of general cargo and their handling equipment are provided in Figures 4.1 and 4.2. It should be borne in mind that the use of the proper tools not only reduces the risks of damage or accidents, but also speeds up the overall cargo operation. The nature of

Figure 4.1 Examples of cargo-handling equipment used for general cargo.

Source: Author's own.

Figure 4.2 Examples of cargo-handling equipment used for general cargo.
Source: Author's own.

the actual cargo and the type of packing used will determine the form of cargo-handling equipment that is employed.

Some common forms of equipment used in the handling of general cargo:

- Broad sling used for the handling of bags of cement.
- Chain slings used in the handling of most steelwork.
- Plate clamps used for the handling of large, flat steel plates.
- Car slings used to lift motor vehicles.
- Heavy lift slings to handle heavy pieces of machinery; for example, a vacuum clamp is used in lifting reels of paper.
- Cargo nets for mailbags and similar cargoes that are not liable to be crushed when hoisted.
- Cargo trays which are ideal for boxes that are easily stacked.

4.4 SAFE HANDLING, STOWAGE, AND CARRIAGE OF TYPICAL GENERAL CARGOES

The safe handling, stowage, and carriage of typical general cargoes can be complex, but in principle, the process should be fairly straightforward. A few examples of how different forms of general cargo may be stowed efficiently are provided herein (Figures 4.3 and 4.4).

Figure 4.3 Stowing sacks.

Source: Bim im Garten, Pexels.

Figure 4.4 Stowing sacks.

Source: Bim im Garten, Pexels.

Bagged cargo is stowed using double dunnage. This is material, frequently timber, which is used to help consolidate and protect the cargo stow, half bag (poor ventilation), or bag and bag (better ventilation). Bagged cargoes must be kept away from the ship's side and bulkheads and must be covered with mats or paper when stowed under the deck head. Bags should be lifted with a canvas sling and handled with dockers' hooks. The bags should be stain-free and not torn. Bags of cement must be stowed compactly to avoid movement. The compartment must be completely dry. The bags are usually lifted with board slings and then manually stowed. Ventilation is critical in most cases of bagged cargo as germination and caking are two common types of damage which are caused through a lack of adequate ventilation.

Cotton is a highly flammable cargo. All fire precautions must be taken when handling cotton. Wet bales of cotton are liable to spontaneous heating and combustion. These must be kept away from the ship's sides and bulkheads. Cotton is also easily damaged by rust.

Rubber is typically shipped in bales wrapped in polythene. These give off a strong odour, which can taint other adjacent cargoes. To avoid this, rubber is normally given top stowage.

Barrels are usually lifted by can hooks if not too heavy and are stowed fore and aft with wedges underneath to prevent free movement. The bung of the barrel must always be facing upwards. Some of the cargo contained in the barrels is liable to damage by taint, for example, wine. To avoid this, the barrels must be stowed in a dark, cool, and dry location.

Drums, as opposed to barrels, should be stowed on their ends. Likely contents include paint, chemicals, and dyes. These may require deck stowage and even ventilation. The drums may contain highly flammable substances. To avoid damaging the drums, always lift using can hooks and stow on wooden dunnage to prevent movement and allow for any leaked substances to flow away.

Cased goods are lifted with rope slings, trays, or, at times, nets. They should be stowed on double dunnage. Heavy cases must be given bottom stowage. If the cargo is pilferable, then lock-up stowage must be used. Cased cargoes may contain goods liable to taint damage or be of an odorous nature. Appropriate precautions should be taken to avoid both from occurring.

Paper reels are normally stowed on end to avoid the risk of distortion and should be well chocked off. Paper is liable to damage by sweat or poor handling. Forklifts with special ends clamp the reels of paper which are then lifted onboard by vacuum clamps. Care should be taken to prevent damage to the edge of the reel.

Steel products range from pig iron to steel billets and pipes. These can severely affect the stability of the vessel. Given their weight and unwieldy nature, steel products are typically lifted by steel chains and stowed at the bottom. They are liable to shift and can be damaged by rust. Some products can be over stowed. Others are liable to overheating and spontaneous combustion such as iron and steel swarf. Coils must be on the round with wedges of dunnage placed underneath. The cargo should be stowed in regular tiers from one side to the other side of the vessel. The stow should be secured as one or more solid blocks. Ensure careful dunnage on long steel products to prevent distortion.

Motor vehicles are always lifted by car slings. If heavy lorries or trucks are to be handled, wire slings are usually attached to the axles' ends. The vehicle must be stowed on a firm level floor with space provided around each vehicle. Ventilation of the compartment to remove carbon monoxide is a factor to consider if cars are wheeled to their stowage position.

Pre-slung goods have the advantage of quick loading and unloading. They can be as large as a 1–2 tonne Intermediate Bulk Container. Over stowage is possible and should be avoided. Pre-slung goods are normally handled with forklifts in the cargo compartment once lifted into the hold by cranes or derricks.

4.5 CARGO HOLD PREPARATION FOR THE RECEPTION OF GENERAL CARGO

The above examples show the diversity in handling and stowage requirements in the general cargo trade. One important factor in the carriage of general cargo is the suitability of the ship's hold or compartments. Generally, a hold which is ready to receive cargo should be clean and dry, well-ventilated, and free from any odours of the previous cargo. To achieve this, the compartment should be prepared accordingly. Clean the stowage space thoroughly, leaving no trace of the previous cargo behind. Do not forget residues which may have coagulated under the hatch covers. If necessary, wash the cargo space down using seawater and allow to dry. When dry, ventilate the space thoroughly. Check with the local Port Authorities for the discharge of wastewater and cargo residues. Remove all rubbish and unwanted dunnage. Clean the bilges thoroughly and check suction. Bilges can be washed with cement or lime. Make sure there are no residual odours. If necessary, coat with bitumastic. Check the fire detection systems and extinguishing equipment. If necessary, blow through the system. Check and clear the scuppers. Check and repair permanent dunnage if necessary. Lay down clean dunnage properly. If double dunnage is used, the lower tier must run athwartships. Check for signs of pests and rodents. It might be necessary to fumigate the holds. Check for guardrails, ropes, and stanchions if in use in the tween decks. Check that the ventilation systems are in good working order. This includes the intakes, exhausts, dampers, and blowers. Check the condition of the tank top, especially around any manholes and ensure there are adequate lashing points for the cargo to be loaded. If necessary, extra supports or rails may need to be welded in place. Check the condition of the hold itself. It should be watertight and should not show signs of excessive wastage around the frames. Lights in the compartment must be in good working condition. They must also be intrinsically safe if the compartment is used to carry dangerous goods or grains as the dusty atmosphere can cause dust explosions.

4.6 UNITISATION

The concept of unitisation is to assist the process of cargo handling by reducing the number of occasions when a piece of cargo needs to be handled, and therefore, also reducing manual handling by mechanical means. Unitised cargoes could be defined as the grouping of two or more items (usually of a homogeneous nature) and securing them with banding, glue, shrink wrap, or slings to form a unit which, together with a base (pallet), allows mechanical handling equipment to lift and transport the unit. Unitisation involves a high measure of palletised cargo. These pallets are loaded by means of wire slings and then stowed in the hold using forklift trucks. The four-way entry pallet is often used in unitised cargo for its convenience. The pallets are stowed on dunnage and often plywood is spread over the bottom tier to level it up and thus create a platform for a new tier. A compact stowage is achievable. The advantages of unitised cargo are handling is reduced to a minimum; there is no damage from cargo hooks; manual handling is negligible; broken stowage is reduced; and loading and discharging rates are improved. When planning, over stowing must always be borne in mind. Damage through contact with break bulk cargo is also a possibility that must not be ignored. The ship officer must also remember that the efficiency of the system relies upon easy access and quick loading or discharging of the cargo. When preparing a compartment to load unitised cargo, the same preparations as are carried out for a compartment to be loaded with general cargo apply.

4.7 CONTAINERISATION

When unitised cargo is mentioned, nothing is more popular than the conventional container that has taken shipping transportation into a new era. Ports, vessels, and road transport revolve around containerisation. Special ships have been built as well as the necessary port installations for the rapid handling of the containers. Standardisation of size has made container handling much easier. However, there are some unusual-sized types which are explained later. Containers are built with their purpose in mind. In some containers, fittings are available from which lashings can be fixed. This way, damage to the goods can be avoided. Sweat is another possible problem that could damage the cargo stacked inside. Since containers are also carried on deck, they are thus subject to varying weather conditions. They tend to be very buoyant if immersed in water, thus securing is an important aspect of the safe carriage of containers. Containers are usually handled by shore gantries or straddle carriers. Because of their heavy weights (usually a TEU can have a mass of up to 24 tonnes), this special equipment is required for their handling. A container is designed to be lifted by fitted attachments at its top four corners. Any other form of lifting imposes strain or may, indeed, cause the container to overbalance. The stowing of containers is usually done in cells in specialised container vessels. Such holds are completely dedicated to either 20- or 40-foot containers and ensure that each succeeding container in a stack rest securely on the weight-bearing corner casting of the one below. The frame is the strongest part of the container, and the side walls are relatively flimsy and prone to damage if handled carelessly. Other conventional vessels would have strengthened structure to be able to support the laden containers. These stronger corner castings would be positioned to fit the bottom corners of the container. In no circumstances should a laden container be stowed outside the dedicated space, as the deck or tank top might not be strong enough to support it. The loading of containers onboard can create many problems, particularly when several different container lengths and port destinations are involved. The vessel must be properly trimmed, and the loading plan must consider dangerous cargo and those containers requiring special attention – for example, reefer container or open top container. Container loading is made easier if the vessel is upright and on an even keel. This is achieved by operating ballast transfer pumps to transfer quantities of water between trimming tanks. An extra margin of safety is required when handling containers because of possible non-standardisation in construction. (See the following extract from *Code of Safe Working Practices for Merchant Seafarers*. Material gratefully reproduced with permission from the UK Maritime and Coastguard Agency).

28.3 *Carriage of containers*

28.3.1 *Containers are simply packages of pre-stowed cargo and sections of Chapter 16, Hatch covers and access lids, and 19, Lifting equipment and operations, may also be relevant to their safe working. Guidance is also published by the UK's Port Skills and Safety organisation in its Health and Safety in Ports series, SIP Leaflet 008 – Guidance on the storage of dry bulk cargo (see the Port Skills and Safety website).*

28.3.2 *Where a container holds dangerous goods, the relevant guidance contained in section 28.2 should be followed.*

For guidance on control of substances hazardous to health, refer to Chapter 21, Hazardous substances and mixtures.

28.3.3 *Freight containers should comply with the International Convention for Safe Containers 1972 (CSC), under which they must carry a safety approval plate (CSC plate). Defective containers, or containers on which the CSC plate is missing, should be reported so that they can be taken out of service. Containers should not be loaded beyond the maximum net weight indicated on the CSC plate and should be in a safe condition for handling and carriage.*

28.3.4 *The equipment used for lifting a container should be suitable for the load, and safely attached to the container. The container should be free to be lifted and should be lifted slowly to guard against the possibility of it swinging or some part of the lifting appliances failing, should the contents be poorly secured, unevenly loaded and poorly distributed or the weight of contents incorrectly declared. The process of loading and securing of goods into a container should follow the IMO/ILO/UN/ECE Guidelines for Packing of Cargo Transport Units (CTUs). Special care should be taken when lifting a container with a centre of gravity that is mobile, for example, a tank container, bulk container or a container with contents that are hanging.*

28.3.5 *Safe means of access to the top of a container should be provided to release lifting gear and to fix lashings. Personnel so engaged should, where appropriate, be protected from falling by use of a properly secured safety harness or other suitable means.*

28.3.6 *All containers should be lashed individually by a competent person. Where containers are stacked, account should be taken of the appropriate strength features of the lashing and stacking-induced stress.*

28.3.7 *On ships not specially constructed or adapted for their carriage, containers should, wherever possible, be stowed fore and aft and securely lashed. Containers should not be stowed on decks or hatches unless it is known that the decks or hatches are of adequate overall and point load-bearing strength. Adequate dunnage should be used.*

28.3.8 *The system of work should be such as to limit the need to work on container tops. Where the design for securing*

containers and checking lashing makes access onto container tops necessary, it should be achieved by means of
the ship's superstructure or by a purpose-designed access
platform or personnel cages using a suitable adapted lifting appliance. If this is not possible, an alternative safe
system of work should be in place.

28.3.9   To allow access to the tops of over-height, soft-top
or tank containers where necessary for securing or
cargo-handling operations, solid top or 'closed containers' should be stowed between them whenever
practicable.

28.3.10 Where the ship's electrical supply is used for refrigerated
containers, the supply cables should be provided with
proper connections for the power circuits and for earthing the container. Before use, the supply cables and connections should be inspected, and any defects repaired
and tested by a competent person. Supply cables should
only be handled when the power is switched off. Where
there is a need to monitor and repair refrigeration units
during the voyage, account should be taken of the need
to provide safe access in a seaway when stowing these
containers.

28.3.11 Personnel should be aware that containers may have
been fumigated at other points in the transport chain,
and there may be a residual hazard from the substances
used.

4.8 PREPARATIONS FOR CONTAINERISED CARGO

4.8.1 Preparing a Cargo Space to Load Containerised General Cargo

In addition to the preparation necessary for break bulk cargo, the following
should be done: (1) check the condition of cell guides, (2) check the availability of lashing equipment (twist locks, bridge fittings etc.), and (3) check
the positioning of the container. This will depend upon the type of container
underneath; for example, no containers should be stacked on an open-
top container. When preparing a space for the reception of containerised
refrigerated cargo, in addition to the steps already mentioned, the following
actions must be taken: (1) check the vessel's electrical power connections,
(2) check all ducted air connections, (3) check that the vessel has spares for
the refrigerated containers, (4) check all temperature recording devices are
fully functional, and (5) ensure there are sufficient power leads of adequate
length for all reefer containers (Figures 4.5 and 4.6).

Figure 4.5 Cellular construction of a typical container ship.

Source: Danny Cornelissen, Magellan Maritime Press Ltd

Figure 4.6 Cellular construction of a typical container ship.

Source: Maersk Line, CC BY 2.0

4.9 EXAMPLES OF COMMON TYPES OF CONTAINERS

General purpose (GP): closed: Suitable for all types of general cargo. With suitable modification, can accommodate solid bulk commodities, granular or powder (Figures 4.7 and 4.8).

Reefer: carry refrigerated cargo: These are fitted with their own refrigeration unit. Require electrical power supply for operation (Figure 4.9).

Bulk container: Carry granular substances or dry powder in bulk. Fitted with openings on the top to allow for loading. Discharge is normally done through a window found on the right-hand door of the container (Figure 4.10).

Ventilated container: Similar to general purpose, but with ventilator galleries along the top and bottom side rails. This allows for passive ventilation of the cargo (Figure 4.11).

Flat rack: Used for odd-shaped cargo which normally extends beyond the dimension of the normal general-purpose container. Cargo can be lashed easily in a flat rack (Figure 4.12).

Figure 4.7 Typical TEU container.

Source: Author's own.

Figure 4.8 Typical FEU container.

Source: Author's own.

Figure 4.9 Typical reefer container.

Source: Author's own.

Figure 4.10 Typical bulk container.

Source: Author's own.

Figure 4.11 Typical ventilated container.

Source: Author's own.

Figure 4.12 Typical flat rack container.

Source: Author's own.

Figure 4.13 Typical tanktainer.

Source: Author's own.

Open top: Again, odd-shaped cargoes. Height of cargo exceeds the height of a normal general-purpose container. There are several securing ports in the floor or along the bottom side rail of the container.

Tank container: Specially designed to carry a specific liquid. Can be a type of dangerous good or a special product (Figure 4.13).

4.10 IDENTIFYING A CONTAINER'S LOCATION

For the easy location of every container onboard, a unique numbering system is used. This method eliminates the risk of mistakes in the handling process. The system employs a six-digit notation, such as 20 06 10
The first two digits represent the BAY number. The middle two digits represent the CELL number, and the last two digits represent the TIER number.

BAY: odd numbers per 20¢ bay
     Even numbers depict 40¢ bay
CELL: numbered odd to starboard and even to port
TIER: under deck evenly from bottom to top on deck evenly commencing
     with 82 from the deck

Figures 4.14–4.20 illustrate the container stowage numbering system.

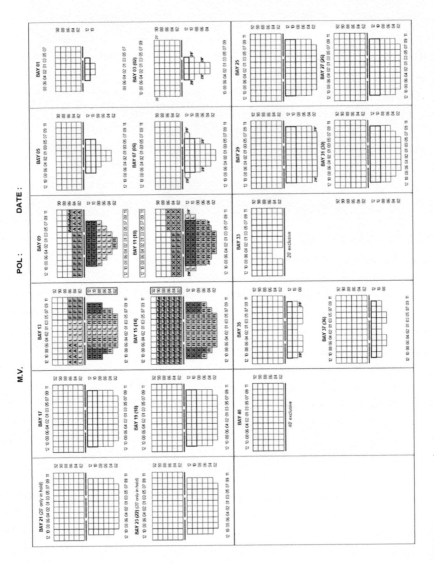

Figure 4.14 Container stowage numbering system.

Source: Author's own.

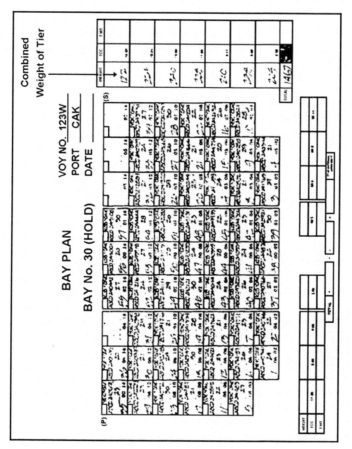

Figure 4.15 Example container stowage plan.

Source: Author's own.

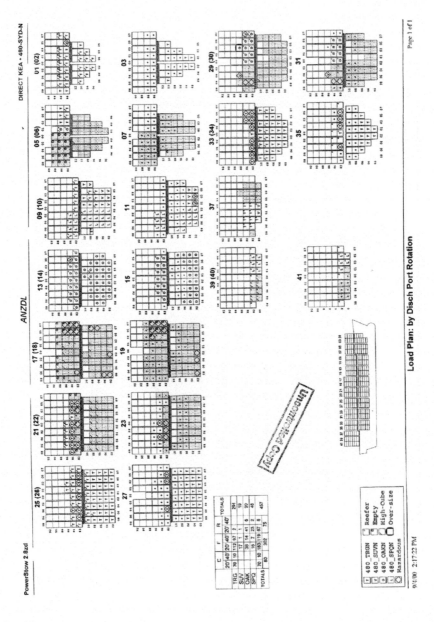

Figure 4.16 Example container stowage plan by load port.

Source: Author's own.

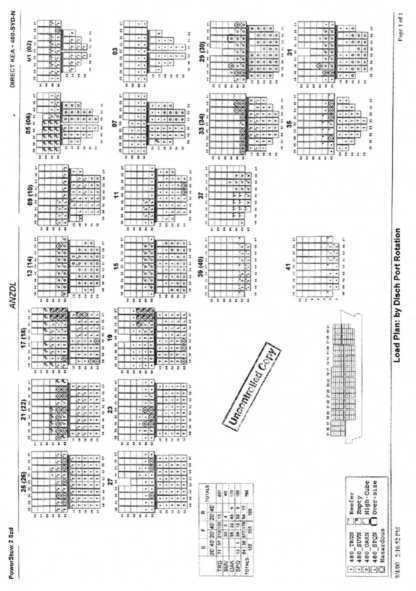

Figure 4.17 Example container stowage plan by discharge port.

Source: Author's own.

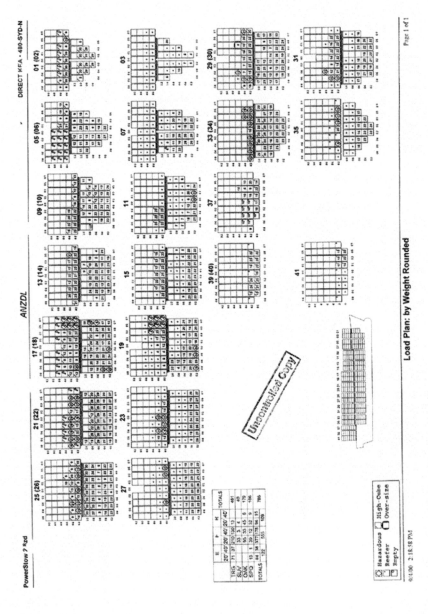

Figure 4.18 Example container stowage plan showing cargo weights.

Source: Author's own.

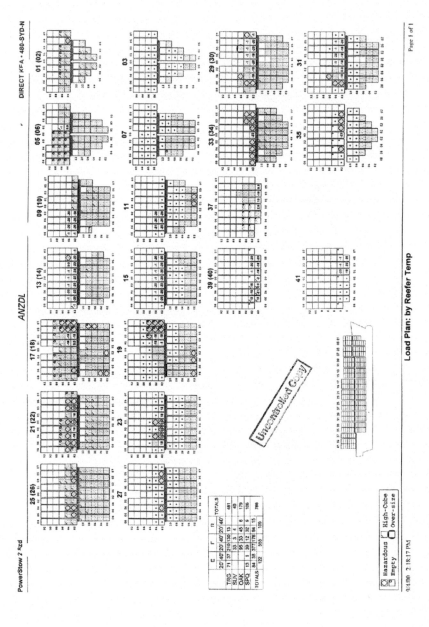

Figure 4.19 Example container stowage plan showing reefer temperature.

Source: Author's own.

010890	010690	010490	010290	010090	010190	010390	010590	010790
010888	010688	010488	010288	010088	010188	010388	010588	010788

Top tier (010888 … 010788):

010888	010688	010488	010288	010088	010188	010388	010588	010788
TRG SYD NOSU4365544 ANZ 8.5 4DR 259 020886	TRG SYD TPHU4928548 ANZ 9.0 4HC 291 020686	TRG SYD TRLU1921054 ANZ 4.2 4RH 291 020486	TRG SYD TOLU6910540 ANZ 4.5 4RH 291 020286	TRG SYD TRLU1783344 ANZ 4.2 4RH 291 020086	TRG SYD TRLU1905583 ANZ 4.2 4RH 291 020186	TRG SYD TRLU1620910 ANZ 4.2 4RH 291 020386	010586	TRG SYD TOLU5910392 ANZ 4.2 4RH 291 020786
TRG SYD CSQU4423948 ANZ 8.4 4DR 259 020884	TRG SYD TRLU4919873 ANZ 30.0 4HC 291 020684	TRG SYD CRXU9148609 ANZ 30.1 4HC 291 020484	TRG SYD CAXU4925907 ANZ 30.3 4HC 291 020284	TRG SYD TRLU5313751 ANZ 30.0 4HC 291 020084	TRG SYD MLCU9801581 ANZ 30.2 4HC 291 020184	TRG SYD CPSU4008085 ANZ 8.0 4DR 259 020384	OXK SYD CRXU2111911 ANZ 12.8 1.3 2DR 259 010584	TRG SYD CRXU9524748 ANZ 8.8 4HC 291 020784

Bay   (02) Above

Figure 4.20 Example bay plan of cargo hold.

Source: Author's own.

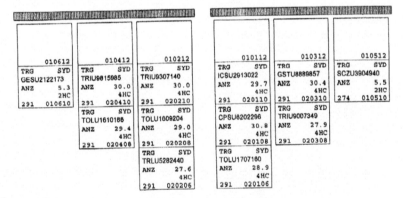

010612	010412	010212		010112	010312	010512
TRG SYD GESU2122173 ANZ 5.3 2HC 291 010610	TRG SYD TRIU9815985 ANZ 30.0 4HC 291 020410	TRG SYD TRIU9307140 ANZ 30.0 4HC 291 020210		TRG SYD ICSU2913022 ANZ 29.7 4HC 291 020110	TRG SYD GSTU8889857 ANZ 30.4 4HC 291 020310	TRG SYD SCZU3904940 ANZ 5.5 2HC 274 010510
	TRG SYD TOLU1610188 ANZ 29.4 4HC 291 020408	TRG SYD TOLU1609204 ANZ 29.0 4HC 291 020208		TRG SYD CPSU6202296 ANZ 30.8 4HC 291 020108	TRG SYD TRIU9007349 ANZ 27.9 4HC 291 020308	
		TRG SYD TRLU5282440 ANZ 27.6 4HC 291 020206		TRG SYD TOLU1707180 ANZ 28.9 4HC 291 020106		

Bay 01 (02) Below

Figure 4.20 (a), (b) Example bay plan of cargo hold.

Source: Author's own.

4.11 REFRIGERATED AND REEFER CARGOES

The success of a 'refrigerated' trade depends principally upon the condition in which the cargo arrives at the port of discharge, and this success is *not* attained solely by the efficiency of the refrigerating machinery or the high standard of the insulation. To ensure that the cargo arrives in the best condition, considerable responsibility rests with cargo officers in the discharge of their duties. It is essential that proper methods of stowage are employed and that all proper precautions are taken both prior to and during the reception of the cargo, also that the cargo is kept at proper temperatures during the voyage. In very broad terms, refrigerated cargoes fall into three main categories:

(1) Frozen cargo.
(2) Chilled cargo.
(3) Temperature-regulated cargo.

4.11.1 Frozen Cargo

Frozen cargo is carried in the hard frozen condition, which means that a temperature of at least –20°C (–4°F) must be attainable. Most frozen commodities are carried at a temperature below –7°C (19.4°F) when no microorganism growth is possible. Some examples are frozen meat, frozen carcasses, and poultry.

4.11.2 Chilled Cargo

Chilled cargo are commodities where the outside has been frozen hard, but the inside remains unfrozen. The carriage of chilled cargo requires some considerable care since the temperature range at which it is carried is normally very small between –2°C to –3°C (from 28.4°F to 26.6°F). Chilled beef and carcasses are examples of chilled cargo. There is also a time limit for cargoes transported in the chilled form as microorganisms still grow at that temperature. Adding CO_2 to the compartment can prolong the lifetime of the chilled cargo.

4.11.3 Temperature-regulated Cargo

Temperature-regulated cargoes are those which are carried at a temperature which restricts processes such as ripening. Some specific commodities require different temperatures, so that goods such as apples can be carried as low as 1°C (33.8°F), whilst citrus fruit such as grapefruit or lemons are carried at 10°C to 12°C (50–53.6°F). In most cases where fruits are carried, quantities of CO_2 are given off (Figure 4.21). The shipper should give

Figure 4.21 Dole chilled cargo container.

Source: Author's own.

guidelines as to the amount of ventilation required, otherwise expert advice should be sought. Vegetables are also carried that way and often cannot be mixed with fruits because of the risks of damage through tainting. Segregation tables are provided by shippers for this reason.

4.12 SAFE HANDLING, STOWING, AND CARRIAGE OF REEFER CARGOES

Essentially, the carriage of refrigerated cargoes (Figures 4.22 and 4.23) could be split into the following steps:

(1) The preparation of the compartment.
(2) The pre-cooling of the compartment.
(3) The loading, which includes the methods of handling and stowage.
(4) The carriage which includes the maintaining of proper temperatures and the precautions to be adopted to prevent undue accumulation of CO_2.
(5) Precautions to be taken during discharge.
(6) Preparing a cargo space for the reception of loose refrigerated cargo.

(7) Space is to be thoroughly cleaned and wiped with a cleansing fluid to prevent the formation of mould.
(8) Space is to be ventilated and deodourised.
(9) Check insulation and repair if necessary.
(10) Bilges to be cleaned sweetened and checked.
(11) Brine traps (they prevent the passage of odour from the bilge to the compartment; they also prevent cold air to travel from the compartment to the bilges) are to be refilled and tested.
(12) Air ducts to be cleaned.
(13) Thermometers to be inspected and checked.
(14) Clean and odourless dunnage must be used.
(15) Gratings, if any, are to be scrubbed.
(16) Ventilators and fixed fire detecting to be plugged for chilled or frozen cargoes.
(17) The space must be pre-cooled before loading.
(18) The space must be surveyed to obtain approval for loading.
(19) If it is necessary to load in a partly filled compartment, then the refrigeration plant must be stopped while loading is going on. Also, the cargo already onboard should be covered with a tarpaulin.

Figure 4.22 Typical refrigerated container.

Source: Andrea Puggioni, CC BY 2.0.

Figure 4.23 'Clip on' refrigeration towers.

Source: Author's own.

4.12.1 Pre-cooling of Compartment

This must be done before loading. Some authorities require the compartment to be pre-cooled up to 48 hours before loading. Temperatures will normally be slightly lower than the normal carrying temperature. However, if ingots of lead, tin, or copper are to be loaded in the compartment, then they can be loaded prior to pre-cooling. If any refrigerated cargo is to be loaded on top of the ingots, then pre-cooling is necessary, and the cargo must not come into contact with the ingots.

4.12.2 Loading and Handling

During loading, officers must ensure that the stow is in such a way that will allow the flow of cold air around the cargo. The loading must be fast to reduce the amount of heat absorbed by the cargo already in the compartment. Cooperation between the engine room and deck is vital as the maintaining of the proper temperatures is of paramount importance. Condensation should not be overestimated when loading general cargo in compartments adjacent to cold compartments. Considerable care and attention are necessary when receiving refrigerated cargo. Coverings and wrappings must be undamaged. Any fruit shipped in an advanced state of ripeness is a prolific source of damage.

4.12.3 CO_2 Control

Ripening fruits evolve heat and CO_2. An excessive amount of CO_2 can cause the fruit to deteriorate. A thermoscope should be used to constantly measure the amount of CO_2 in the compartment. If found in excess, then the CO_2 must be extracted from the compartment.

4.12.4 Methods of Transportation of Reefer Cargoes

Refrigerated cargo is transported at sea by two main methods:

(1) Loose or unitised refrigerated cargo transported aboard custom-built reefer ships or traditional type general cargo vessels with limited reefer capacity.
(2) Containerised reefer cargo transported aboard container vessels which have full or partial refrigerated capacity.

4.12.5 Loose and Unitised Reefer Cargoes

Onboard reefer ships or in reefer compartments, the cargo is loaded manually or by forklift trucks. The cargo temperature is usually controlled by a forced cold air ventilation system. The refrigerated chambers have temperature-monitoring devices that can be coupled to a remote temperature control device which automatically actuates or stops the cooler room's fans, thus maintaining the desired carrying temperature.

4.13 CONTAINERISED REEFER CARGOES

The carriage of refrigerated cargoes in containers falls into two broad categories: ship-dependent containers and independent containers.

4.13.1 Ship-dependent Containers

Ship-dependent containers require power from the vessel as the source of energy. They have their own refrigeration plant and temperature monitoring devices. Another type of container, which is also ship-dependent, is the type that requires cold air to be blown inside the container. The cold air is provided by the ship's refrigeration plant.

4.13.2 Independent Containers

Independent containers have their own source of power and refrigeration plant. The need for ventilation (because of their internal combustion engine) makes them unsuitable for under deck carriage.

Refrigerated cargoes are generally high-value cargoes, and as such attract high freight rates. Because of this, any failure by ship's staff to care for the cargo in a correct manner can be an extremely costly exercise. The maintenance of correct carrying temperatures cannot be overemphasised, as this is the prime responsibility of ship staff so far as on-board care of refrigerated cargo is concerned.

In this chapter, we have examined some of the key concepts around general cargo stowage, unitised cargo and containerised cargo. We have also briefly looked at how containers are stowed with the ship's holds. Needless to say, containerisation has greatly improved the dry bulk trade, so much so that general cargo ships are increasingly a rarity. Even so, it is useful for the modern seafarer to have at least a minimum understanding of how these ships function. In the next chapter, we will look at RORO cargo, and how rolling stock is loaded, stowed, and discharged.

Chapter 5

Principles for the Safe Handling, Stowage, and Carriage of RORO Cargo

5.1 INTRODUCTION

ROROs are designed to convey road haulage vehicles, cars, and goods secured on road trailers (Figures 5.1 and 5.2). The cargo is loaded onto trailers, which are then wheeled in and out of the huge cargo compartment. Access to the compartment is through stern-mounted, side-mounted, or bow-mounted ramps. The main objective of RORO operations is very fast turnaround. For optimum operational efficiency, good planning and proper supervision are of paramount importance. The loading time of a RORO vessel is often dependent upon the time taken to manoeuvre and secure the cargoes, trailers, and vehicles. The factors which affect the efficiency of RORO operations include (but are not limited to) the following:

(1) The gradient of each ramp.
(2) The width of each ramp.
(3) Any bends and turns to be negotiated, and blind corners.
(4) The speed of operation of elevators or other similar handling equipment.
(5) The 'vehicle envelope' (i.e. the overhead clearance) of the cargo and the deck above or hanging infrastructure such as pipework.
(6) Any changes in deck or ramp gradient.
(7) The organisation and flow of traffic.

Thorough knowledge of the port's tidal conditions is important. Thus, ballasting could be used to good advantage to reach and maintain an optimum ramp gradient.

DOI: 10.1201/9781003354338-6

Figure 5.1 Example of RORO cargo stowage.

Source: Author's own.

Figure 5.2 Example of RORO cargo stowage.

Source: Author's own.

5.2 RORO CARGO OPERATIONS

5.2.1 Rolling Cargo

The stowage of RORO cargo onboard depends on the way in which the cargo is brought onboard. This maybe in any of the following ways: road vehicles with an integral prime mover or haulage power which will also remain with the vessel; road trailers which will remain with the ship throughout the sea passage; roller trailers which are not suitable for road haulage; cargo towed onboard using roller trailers which are then removed and stowed without their wheels; cargo secured on flatbeds and carried onboard either using roller trailers or by some other mechanical handling equipment, with both the flatbed and its cargo being stowed as a single unit; pallets either singularly or in groups carried onboard using roller trailers or forklift trucks; and, individual items of cargo brought onboard by forklift trucks. Another potential source of concern is the securing of the cargo and its trailer. The ships' personnel must ensure that the cargo is properly secured to the trailer as a first measure. Then the securing of the trailer can be carried out. Securing of cargo will be dealt with in another section.

5.2.2 Vehicles

Vehicles are usually parked close together in lanes which are approximately 3 metres (9.8 ft) wide. This allows the lashing gangs to secure each vehicle properly. Trailers may be backed up the ramp so that, at the port of discharge, they are simply towed away without the need to turn around. The need for proper ventilation is emphasised when mechanical equipment, towing vehicles, and road vehicles are operating on vehicle decks.

5.2.3 Containers

Containers can be stowed fore and aft or athwartships. If reefer containers with integral refrigeration units are carried under deck, adequate ventilation must be present.

The following are some of the points that the prudent officer is well advised to consider when handling RORO cargoes:

- The vehicle and/or load is similarly secured and appropriately fitted.
- The decks of the loading areas are not wet, greasy, or slippery.
- Access to vulnerable parts within the loading area is not obstructed (doors, drainage outlets, LSA/FFA).
- With particularly heavy vehicles, it may well be necessary to add to the normal securing devices jacking arrangements or other frictional resistance material placed beneath the unit, onto the deck.

- Vehicles carrying dangerous goods must be segregated from other cargoes.
- Vehicles with flats, open-sided bogies, or trailers must be so loaded and secured that the contained cargoes should not move, or slip, against the movement of the vessel.
- Fore and aft lane stowage is the more desirable procedure with vehicles, with brakes on and the engine on gear.
- High vehicles which are loaded and with a relatively high centre of gravity require a particular degree of security and lashing provisions.

5.3 GUIDANCE FOR CARGO OFFICERS OVERSEEING RORO OPERATIONS

The following guidance is extracted from the *Code of Safe Working Practices for Merchant Seafarers*, which is reproduced with kind permission of the UK Maritime and Coastguard Agency.

5.3.1 General Precautions

The movement, stowage, and securing of vehicles on vehicle decks and ramps should be supervised by a responsible ship's officer, assisted by at least one competent person. Smoking and naked flames should not be permitted on any vehicle decks. Conspicuous 'no smoking' or 'no smoking/naked lights' signs should be displayed (in accordance with MGN 341(M), there should be no unauthorised persons on vehicle decks at any time, and there should be no entry to vehicle decks when the vessel is at sea, unless specifically permitted. Passengers and drivers should not be permitted to remain on vehicle decks without the express authority of a responsible ship's officer. The period prior to disembarkation when passengers and drivers are requested to return to their vehicles should be kept to a minimum. Where closed-circuit television (CCTV) cameras are fitted, they should, where practicable, have an uninterrupted view of the vehicle deck. The use of CCTV for continuous watch does not necessarily preclude the need for car-deck patrols, for example, coupled with fire patrols of passenger accommodation.

5.3.2 Ventilation

Vehicle decks should have adequate ventilation at all times, with special regard to hazardous substances. In accordance with S.I. 1998/1011 and S.I. 1998/1012, on passenger vessels, ventilation fans in closed RORO spaces must normally be run continuously whenever vehicles are onboard. An increased number of air changes may be required when vehicles are being loaded or unloaded, or where flammable gases or liquids are stowed in a closed RORO space. Merchant shipping regulations specify the special requirements for cargo space ventilation. To reduce the accumulation of fumes, drivers should be instructed to stop their engines as soon as practicable after embarking and to avoid starting up prior to departure until instructed to do so. During loading and discharging, ventilation may be improved by keeping both bow and stern doors open, provided there is adequate freeboard at these openings. When there is doubt about the freshness of the atmosphere, arrangements should be made for testing to ensure the maintenance of 20% oxygen and a carbon monoxide content below 30 ppm in the atmosphere of the space.

5.3.3 Fire Safety and Prevention

Fire-detection systems should be switched on whenever vehicle decks are unattended. Deck and engine crew should be trained in the use of the drencher systems and their operation. Continuous monitoring of vehicle decks by CCTV or regular fire patrols should also be in place. All fire doors should be kept closed on vehicle decks when the vessel is at sea.

5.3.4 Noise

Personnel working on vehicle decks should not be exposed to the equivalent of 85 dB(A) or greater when averaged over an 8-hour day. Hearing protection should be available for use when the noise level is equivalent to or exceeds 80 dB(A) averaged over an 8-hour day and should be worn when it is equivalent to or exceeds 85 dB(A) averaged over an 8-hour day. For further guidance on noise levels, see Chapter 12, Noise, vibration and other physical agents, of this Code and the *Code of Practice for Controlling Risks due to Noise on Ships* (revised 2009).

5.3.5 Safe Movement

Pedestrians should be warned of vehicle movements when entering or crossing car or vehicle decks and keep to walkways when moving about the ship. As far as possible, routes used by vehicles should be separated from pedestrian passageways, and the use of ship's ramps for pedestrian access should be avoided. Ramps that are used by vehicles should not be used for pedestrian access unless there is suitable segregation of vehicles and pedestrians. Segregation can be achieved through the provision of a suitably protected walkway, or by ensuring that pedestrians and vehicles do not use the ramp at the same time (see the Code of Practice on the Stowage and Securing of Vehicles on RORO vessels refer to Code of Safe Working Practices for Merchant Seafarers (COSWP) section 2.6). Crew members should exercise great care when supervising the driving, marshalling, and stowing of vehicles to ensure that no person is put at risk. The following precautions should be taken: crew should be easily identifiable by passengers. Personnel required to be on the vehicle decks should wear appropriate personal protective equipment, including high-visibility clothing. Communications between deck officers and ratings should be clear and concise to maintain the safety of passengers and vehicles. There should be suitable traffic-control arrangements, including speed limits and, where appropriate, the use of signallers. Collaboration may be necessary with shore-side management where they also control vehicle movements onboard ship. Hand signals used by loading supervisors and personnel directing vehicles should be unambiguous. Adequate illumination should be provided. Personnel directing vehicles should keep out of the way of moving vehicles, particularly those that are reversing, by standing to the side, and where possible should remain within the driver's line of sight. Extra care should be taken at the 'ends' of the deck where vehicles may converge from both sides of the ship. Crew members should be wary that vehicles may lose control on ramps and sloping decks, especially when wet, and that vehicles on ramps with steep inclines may be susceptible to damage. Ramps should have a suitable slip-resistant surface. Where fitted, audible alarms should be sounded by vehicles that are reversing. Safe systems of work should be provided in order to ensure that all vehicle movements are directed by a competent person.

5.3.6 Use of Work Equipment

Ships' ramps, car platforms, retractable car decks, and similar equipment should be operated only by competent persons authorised by a responsible ship's officer, in accordance with the Company's work instructions. Safe systems of work should be provided to ensure that the health and safety of crew or passengers is not put at risk. Ramps, etc. should not be operated unless the deck and ramp can be seen to be clear of people and if any person appears on the deck while the ramp is moving, the operation should be stopped immediately. Where possible, such ramps and decks should be fitted with audio and visual alarms. Training in the use of such equipment should consist of theoretical instruction enabling the trainee to appreciate the factors affecting the safe operation of the plant, and supervised practical work. Moveable deck ramps should be kept clear of passengers when being raised or lowered. When cars are lowered on the ramps of moveable decks, they should be suitably chocked. If the operator cannot clearly see the whole operation from the control station, then a lookout should be posted to ensure ramp and landing areas remain clear throughout the operation. No person should be lifted by ramps, retractable car decks, or lifting appliances, except where the equipment has been designed or especially adapted for that purpose. Retractable car decks and lifting appliances should be securely locked when in the stowed position. After all vehicles have been loaded, the car-deck hydraulics should be isolated so that they cannot be accidentally activated during the voyage, and the bridge should be informed.

The ship's mobile-handling equipment, which is not fixed to the ship, should be secured in its stowage position before the ship proceeds to sea.

5.3.7 Inspection of Vehicles

Before being accepted for shipment, every freight vehicle should be inspected externally by a competent and responsible person or persons to check that it is in a satisfactory condition for shipment, for example: it is suitable for securing to the ship in accordance with the approved cargo-securing manual (see also section 28.1.4); where practicable, the load is secured to the vehicle; the deck or doorway is high enough for vehicles to pass through and vehicles have adequate clearance for ramps with steep inclines; and any labels, placards, and marks that would indicate the carriage of dangerous goods are properly displayed. It is important to ensure, so far as is reasonably practicable, that on each vehicle the fuel tank is not so full as to create a possibility of spillage. No vehicle showing visual signs of an overfilled tank should be loaded. In accordance with MGN 341(M), seafarers should be aware of hazardous units as detailed on the stowage plan and indicated by labels, placards, and marks, and should be on guard against the carriage of undeclared dangerous goods.

5.3.8 Stowage

Shippers' special advice or guidelines regarding handling and stowage of individual vehicles should be observed. Vehicles should, so far as possible, be aligned in a fore and aft direction; be closely stowed athwartships so that in the event of any failure in the securing arrangements or from any other cause, the transverse movement is restricted. However, sufficient distance should be provided between vehicles to permit safe access for the crew and for passengers getting into and out of vehicles and going to and from accesses serving vehicle spaces; and be so loaded that there are no excessive lists or trims likely to cause damage to the vessel or shore structures. Vehicles should not be parked on permanent walkways; be parked so as to obstruct the operating controls of bow and stern doors, entrances to accommodation spaces, ladders, stairways, companionways or access hatches, firefighting equipment, controls to deck scupper valves or controls to fire dampers in ventilation trunks; or be stowed across water spray fire curtains, if these are installed. Safe means of access to securing arrangements, safety equipment, and operational controls should be properly maintained. Stairways and escape routes from spaces below the vehicle deck should be clearly marked with yellow paint and kept free from obstruction at all times. The parking brakes of each vehicle or each element of a vehicle, where provided, should be applied and the vehicle should, where possible, be left in gear. Semi-trailers should not be supported on their landing legs during sea transport unless the landing legs are specially designed for that purpose and so marked, and the deck plating has adequate strength for the point loadings. Uncoupled semi-trailers should be supported by trestles or similar devices placed in the immediate area of the drawplates so that the connection of the fifth wheel to the king-pin is not restricted. Drums, canisters, and similar thin-walled packaging are susceptible to damage if vehicles break adrift in adverse weather and should not be stowed on the vehicle deck without adequate protection. Depending on the area of operation, the predominant weather conditions and the characteristics of the ship, freight vehicles should be stowed so that the chassis are kept as static as possible by not allowing free play in the suspension. This can be done by securing the vehicles to the deck as tightly as the lashing tensioning device will permit. Care should be taken to ensure lashings are not over-tightened. Only designed tensioning arrangements should be used and no additional extensions should be used to increase tightening force. Alternatively, the freight vehicle chassis may be jacked up prior to securing. Because compressed air suspension systems may lose air, adequate arrangements should be made to prevent the slackening off of lashings as a result of air leakage during the voyage. Such arrangements may include the jacking up of a vehicle or the release of air from the suspension system where this facility is provided.

5.3.9 Securing of Cargo

Securing operations should be completed before the ship proceeds to sea. Within the constraints laid down in the approved cargo-securing manual, the master has the authority to decide on the application of securings and lashings and the suitability of the vehicles to be carried. In making this decision, due regard shall be given to the principles of good seamanship, experience in stowage, good practice and the IMO Code for Cargo Stowage and Securing (CSS Code). Seafarers appointed to carry out the task of securing vehicles should be trained in the use of the equipment to be used and in the most effective methods for securing different types of vehicles. Securing operations should be supervised by competent persons who are conversant with the contents of the cargo-securing manual. Freight vehicles of more than 3.5 tonnes should be secured in all circumstances where the expected conditions for the intended voyage are such that movement of the vehicles relative to the ship could be expected. During the voyage, the lashings should be regularly inspected to ensure that vehicles remain safely secured. Seafarers inspecting vehicle spaces during a voyage should exercise caution to avoid being injured by moving or swaying vehicles. If necessary, the ship's course should be altered to reduce movement or dangerous sway when lashings are being adjusted. The officer of the watch (OOW) should always be notified when an inspection of the vehicle deck is being made. When wheel chocks are being used to restrain a semi-trailer, they should remain in place until the semi-trailer is properly secured to the semi-trailer towing vehicle. No attempt should be made to secure a vehicle until it is parked, the brakes (where applicable) have been applied and the engine has been switched off. When vehicles are being stowed on an inclined deck, the wheels should be chocked before lashing commences. The tug driver should not leave the cab to disconnect or connect the trailer brake lines. A second person should do this. The parking brake on the tug should be engaged and in good working condition. As well as wheel chocks, at least two lashings holding the unit against the incline should be left in place until the trailer's braking system is charged and operating correctly. Where seafarers are working in shadow areas or have to go under vehicles to secure lashings, hand lamps and torches should be available for use. Seafarers engaged in the securing of vehicles should take care to avoid injury from projections on the underside of the vehicles. An agreed method of signalling between the driver and the lashing crew should be established, preferably by the use of a whistle or other distinct sound signal.

Wherever possible, lashings should be attached to specially designed securing points on vehicles, and only one lashing should be attached to any one aperture, loop, or lashing ring at each securing point. When tightening lashings, care should be exercised to ensure that they are securely attached to the deck and to the securing points of the vehicle. Hooks and other devices that are used for attaching a lashing to a securing point should be applied in a manner that prevents them from becoming detached if the lashing slackens during the voyage. Lashings should be so attached that, provided there is safe access, it is possible to tighten them if they become slack. Lashings on a vehicle should be under equal tension. Where practicable, the arrangement of lashings on both sides of a vehicle should be the same, and angled to provide some fore and aft restraint, with an equal number pulling forward as are pulling aft. The lashings are most effective on a vehicle when they make an angle with the deck of between 30 degrees and 60 degrees. When these optimum angles cannot be achieved, additional lashings may be required. Where practicable, crossed lashings should not be used for securing freight vehicles because this arrangement provides no restraint against tipping over at moderate angles of roll of the ship. Lashings should pass from a securing point on the vehicle to a deck securing point adjacent to the same side of the vehicle. Where there is concern about the possibility of low coefficients of friction on vehicles such as solid-wheeled trailers, additional crossed lashings may be used to restrain sliding. The use of rubber mats should be considered. Lashings should not be released for unloading, before the ship is secured at the berth, without the master's express permission.

Seafarers should release lashings with care to reduce the risk of injury when the tension is released. To avoid damage during loading and unloading, all unused securing equipment should be kept clear of moving vehicles on the vehicle deck. A competent person should inspect securing equipment to ensure that it is in sound condition at least once every six months and on any occasion when it is suspected that lashings have experienced loads above those predicted for the voyage. Defective equipment should be taken out of service immediately and disposed of or placed where it cannot be used inadvertently. Unused lashing equipment should be securely stowed away from the vehicle deck.

5.3.10 Dangerous Goods

This section should be read in conjunction with COSWP chapter 21, Hazardous Substances and Mixtures. For guidance on dealing with emergencies involving dangerous goods, see COSWP chapter 4, Emergency Drills and Procedures, and the International Maritime Dangerous Goods (IMDG) Code. Prior to loading, freight vehicles carrying dangerous goods should be examined externally for damage and signs of leakage or shifting of contents. Any freight vehicle found to be damaged, leaking or with shifting contents should not be accepted for shipment. If a freight vehicle is found to be leaking after loading, a ship's officer should be informed and personnel kept well clear until it is ascertained that no danger to personnel persists. Freight vehicles carrying dangerous goods and adjacent vehicles should always be secured. Tank vehicles and tank containers on flat-bed trailers containing products declared as dangerous goods should be given special attention. Pre-voyage booking procedures should ascertain those tanks have been approved for the carriage of their contents by sea.

5.3.11 Specialised Vehicles

Gas cylinders used for the operation and business of vehicles such as caravans should be adequately secured against movement of the ship, with the gas supply cut off for the duration of the voyage. Leaking and inadequately secured or connected cylinders should be refused for shipment. In accordance with MGN 341(M) MGN 545(M+F) MGN 552(M), the following vehicles, trailers, and loads should be given special consideration: tank vehicles or tank containers containing liquids not classified as dangerous goods. These may be sensitive to penetration damage and may act as a lubricant. These vehicles must always be secured. Tracked vehicles and other loads making metal-to-metal contact with the deck; where possible, rubber mats or dunnage should be used. Loads on flat-bed trailers. Vehicles with hanging loads, such as chilled meat or floated glass.

5.3.12 Partially Filled Tank Vehicles

Freight vehicles carrying livestock require special attention to ensure that they are properly secured, adequately ventilated, and stowed so that access to the animals is possible. Further guidance is contained in the Department of the Environment, Food and Rural Affairs (Defra) Regulation on the Welfare of Animals During Transport: New rules for transporting animals (see Appendix 2, Other sources of information). Where vehicles are connected to electrical plug-in facilities, personnel should take the appropriate precautions as described in COSWP chapter 18, Provision, care and use of work equipment, of this Code for working with any electrical equipment.

5.3.13 Housekeeping

All walkways should be kept clear. All vehicle decks, ships' ramps, and lifting appliances should, so far as is reasonably practicable, be kept free of water, oil, grease, or any liquid that might cause a person to slip or that might act as a lubricant to a shifting load. Any spillage of such liquid should be quickly cleaned up; sand boxes, drip trays, and mopping-up equipment should be available for use on each vehicle deck. All vehicle decks, ships' ramps, and lifting appliances should be kept free of obstructions and loose items such as stores and refuse. Seafarers should be careful to avoid electrical points and fittings when washing down vehicle decks. All scuppers should be kept clear of lashing equipment, dunnage, etc.

Because of the nature of the RORO vessel (with huge areas prone to free surface effect) and the increased risks of cargo shifting, cargo operations must be closely supervised by the officer. History, through the sinking of the *MS Herald of Free Enterprise* (1987) and the *MS Estonia* (1994), still haunts the mind of seafarers and highlights the dangers associated with the RORO (Figures 5.3 and 5.4).

Figure 5.3 MS Herald of Free Enterprise (1987).

Source: Author's own, Open Government Licence.

Figure 5.4 MS Estonia (1994) (scale model).

Source: Author's own.

Chapter 6

Principles for the Safe Handling, Stowage, and Carriage of Dry Bulk Cargo

6.1 INTRODUCTION

Solid bulk cargoes comprise iron and other ores, coal, grain, bauxite, and phosphate among many others. These bulk cargoes are usually shipped in specially designed bulk carriers or bulkers and are loaded directly into the ship with no other containment or packaging. Bulkers normally are full form with large clear holds to aid loading and discharge. Refer to Figure 6.1.

Their carriage is governed by the *Code of Safe Practice for Solid Bulk Cargoes*, the purpose of which is to assist persons responsible for the safe stowage and shipment of bulk cargoes. It is published and updated regularly by the IMO. Solid bulk cargoes are dangerous because of their toxicity, flammability, and ability to shift. The hazards involved with the carriage of solid bulk cargoes, together with suggestions to counteract these, are outlined in the Code. When solid bulk cargoes are carried, the following considerations should be borne in mind:

- Weight distribution of the cargo (usually small stowage factor).
- Stability of the vessel in all sea and load conditions.
- The nature of the cargo.
- The properties of the cargo, and any potential chemical reactions.
- Any information available from the shipper, including Material Safety Data Sheets (MSDS).

The *Code of Safe Practice for Solid Bulk Cargoes* is divided into two sections:

(1) List of cargoes which may liquefy.
(2) List of bulk materials possessing chemical hazards.

With respect to cargoes which may liquefy, their moisture content and their transportable moisture limit must be ascertained from the shipper before loading and assessing the stability of the vessel. Because of their potential to shift and affect the vessel's stability, extreme care should be

DOI: 10.1201/9781003354338-7

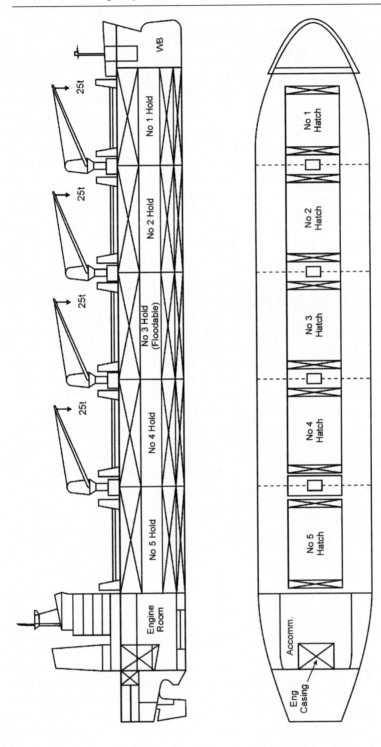

Figure 6.1 Typical bulk carrier layout.

Source: Author's own.

exercised throughout the loading and stowing process. The UK MCA issued guidance on 1 March 2004 which implemented EC Directive 2001/96/EC in the document *Safe Loading and Unloading of Bulk Carriers, 2003*. In addition, the Australian Maritime Safety Authority has also published the *Australian Manual of Safe Loading, Ocean Transport and Discharge Practices for Dry Bulk Commodities*, an extract from which is included in the following sections. The information in this manual aims to assist cargo officers in obtaining information for most solid bulk cargoes.

6.2 THE LOADING AND HANDLING OF DRY BULK CARGOES

Bulk cargoes are usually loaded from a sprout or tip. Conveyor belts are also quite common. Loading rates exceeding 4,500 tonnes per hour are achievable at many ports around the world. The loading procedures involves specific attention being given to the following:

(1) The density of the cargo.
(2) The moisture content of the cargo.
(3) Temperature requirements during the passage.
(4) The angle of repose of the cargo (Figure 6.2).

As an example, refer to the following extract from the *Australian Manual of Safe Loading, Ocean Transport and Discharge Practices for Dry Bulk Commodities* regarding the carriage of Fly Ash. Cargoes with a low angle of repose are liable to shift. This means the correct loading procedures must be adhered to. Cargoes that are liable to give off some of their moisture content must be stowed in compartments where bilges are cleaned and pumpable. Trimming of the cargo must be carried out as much as possible. Watertight integrity must be maintained. Personnel must be prohibited from entering any compartment unless it has been thoroughly ventilated. Cargo discharge is carried out by grabs

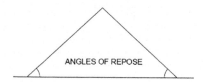

Figure 6.2 Angle of repose.

Source: Author's own.

or elevators (such as suction tubes). Bulldozers are usually used within the compartment to facilitate efficient cargo discharge, though these have been known to cause severe damage to vessels at times. Their presence onboard must be closely monitored. The digging rate of unloaders can reach up to 2,500 tonnes per hour (tph).

Extract from the *Code of Safe Working Practices for Australian Seafarers*, which has been reproduced with permission from the Australian Maritime Safety Authority:

24.2 Bulk carriers and the carriage of bulk cargoes
>   *Note:* This section should be read in conjunction with the IMO Bulk Cargoes Code.

24.2.1   The dust created by certain cargoes, particularly in loading, discharging or hold cleaning operations, may pose an explosion hazard and should be limited as far as possible to the minimum.

24.2.2   Many solid bulk cargoes, some seemingly innocuous, can cause health problems for seafarers in various ways. For example:
(a) ammonium nitrate fertilisers produce toxic gases upon decomposition;
(b) antimony ore dust is toxic;
(c) barium nitrate dust on food is toxic if swallowed;
(d) pencil pitch when handled may cause severe irritation of the skin and eyes in sunlight.

24.2.3   Portholes, doors etc., should be kept closed in port if they permit cargo dust to enter the ship's accommodation area.

24.2.4   Spaces used for the carriage of bulk cargoes should be treated as confused or dangerous spaces. The procedures for entering such spaces, set out in chapter 10, should be strictly followed.

24.2.5   The properties of dry bulk cargoes should be carefully considered as certain bulk materials are liable to oxidation. This may result in oxygen reduction, emission of toxic fumes and self-heating. Other materials may emit toxic fumes, particularly when wet. Other materials, if they become wet, are corrosive to skin, eyes and mucous membranes, as well as to the ship's structure.

24.2.6   Ships which carry cargoes that may emit toxic gases, for whatever reason, should be provided with the appropriate gas detection equipment.

24.2.7 Many bulk cargoes, particularly ores, are loaded into holds from great heights and at very fast rates. This can create sufficient stress to damage the structure of the vessel. This could be avoided by reducing the loading rate.

24.2.8 The plans for the loading and discharging of ships should be properly adhered to, so that the vessel is not exposed to unacceptable stresses, shear forces and bending moments. There is a SOLAS requirement to develop and execute a formal loading plan. The Bulk Cargoes Code contains details. See also the *IMO Code of Practice for Safe Loading and Unloading of Bulk Carriers* (the BLU Code).

24.2.9 Some kinds of cargoes, including concentrates, certain coals and other materials with similar physical properties, may liquefy above the transportable moisture limit and cause a shift in cargo. The moisture content should therefore be carefully checked prior to loading and during the voyage, as moisture migration may occur due to vibration and/or ship motion.

24.2.10 Precautions to prevent liquids entering through pipelines into cargo holds in which bulk cargoes are stowed should be maintained throughout the voyage.

24.2.11 Precautions should be taken against seawater entering holds through hatch covers moving or flexing when the ship is working in a seaway.

24.2.12 Water should not be used to cool cargoes that may liquefy.

24.2.13 The appropriate national and international requirements with respect to ventilation should be observed. Certain cargoes, such as some coals, copra, swarf, charcoal and concentrates etc., are liable to self-heating and may catch fire if the temperature is high enough. Cooling such material should be carried out with extreme care since water used to cool the cargo may actually cause increased heating and lead to spontaneous ignition and/or explosion. The temperature of holds containing such cargo should be checked daily or in accordance with the shipowner's or shipper's instructions.

24.2.14 The dust from some bulk cargoes including grain and sugar dusts can be explosive. Particularly when cleaning holds after discharge, seafarers should be made aware of this hazard. Smoking should be prohibited or restricted and cleaning carried out so as to minimise dust formation, for example, by hosing down. Static electricity is a major source of hazard and care must be taken to ensure that equipment used is suitable for controlling static hazards.

24.2.15 Employees should not enter wing tanks when grain is being loaded.

6.3 PRINCIPLES AND PRACTICE FOR THE SAFE HANDLING, STOWAGE, AND CARRIAGE OF SOME SPECIAL TYPES OF SOLID BULK CARGO CONCENTRATES

Solid bulk cargo concentrates are usually powdery in character and carry considerable moisture. The moisture content may have seriously increased should there have been heavy rain during stowage or loading, or leakage into the hold during the voyage. All concentrates are liable to shift during the voyage. Precautions such as erection of shifting boards must be taken so as to minimise the effect of cargo shifting. Over-stowing could be an alternative to shifting board.

- *Ores:* Ores usually have a small stowage factor and, therefore, usually cause the vessel to be rigid. The structural soundness of the vessel is usually put to the test when ores are loaded. It is often loaded in alternate holds, untrimmed if the ship is designed for such loading and the stresses have been checked carefully. Finely crushed ores are liable to absorb a considerable amount of moisture. Vibration of the vessel can cause the moisture-saturated ore to become slurry and create a list. Bagging has been found to be necessary to prevent the cargo from shifting.
- *Coal:* The carriage of coal is hazardous because of the evolution of methane gas, also known as marsh gas. Mixed in the right proportion with air, methane can produce huge explosions. Apart from the risks of explosion, coal is also liable to spontaneous combustion. The amount of heating that takes place depends upon the type of coal loaded. To reduce the risk of explosion, surface ventilation is recommended to remove the methane lying on top of the cargo. Ventilation into a mass of coal is dangerous. To reduce the risk of spontaneous combustion, coal should be kept as cool as possible and free from through ventilation. When loading coal, every effort should be made to reduce breakages. This affects the stowage pattern with powdery coal in the centre and large pieces of coal down the sides. The large pieces of coal allow for good air circulation to the powdery coal, thus presenting the risks of spontaneous combustion. Spar ceiling must be removed to reduce the possibility of air pockets. Refer to House, D., *Cargo Work for Maritime Operations*, 7th edition (Elsevier, London) for more on ventilation for coal cargo.

6.4 PREPARING A SPACE FOR THE RECEPTION OF DRY BULK CARGO

Though very similar to the preparation involved in loading general cargo, the loading of dry bulk will necessitate some additional and particular attention (Table 6.1):

(1) Removal of unnecessary dunnage (spar ceiling).
(2) Bilges and drains to be covered with burlap to prevent the cargo to run into the bilges but still allow water to run through.
(3) Inspect hold for damage as per the following diagram.
(4) Sometimes it may be necessary to erect shifting boards to counteract the effect of cargo shifting. The authorities must approve this.
(5) Prevent dust from entering deck machinery.
(6) Check sounding pipes, air pipes, and ventilators.

During the cargo loading operation, cargo officers are strongly advised to encourage stevedores to discharge cargo sweepings as far as possible, ensure crew or stevedores sweep down the deck heads and bulkheads as the discharge proceeds. This is particularly relevant to grain cargoes. If cargo has filled a bilge, direct the crew to clean it prior to the completion of the discharge, so the contents can be discharged efficiently. Sweep the holds before washing to remove any bulky cargo residues. Holds should not be washed down where the same cargo is to be carried again and the charterers do not want the holds to be cleaned; or clean cargo such

Table 6.1 Locations and Indications of Cracks and Deformations

Where to Look	What to Look for
Side shell plating	Cracks in welds or plates
	Leaks in welds or plates
	Distortion of plating
Connection of bulkhead plating to side shell	Punctured plating
	Cracked plating
	Heavily indented plating
	Buckling plating
	Corrosion and wastage
Connection of side shell frames and end brackets to the shell plating and hopperside tank plating by close-up inspection	Cracks
	Corrosion and wastage
	Excessively deformed frames and brackets
	Detached frames and brackets
Connection of side shell frames and end brackets to the shell plating and topside tank plating	Cracks
	Corrosion and wastage
	Excessively deformed frames and brackets
	Detached frames and brackets

as steel coils are to be carried; or, freezing conditions do not permit the washing down of the hold; or, where the vessel will remain in areas where the discharge of washing is not permitted. If hold washing is to be carried out, first decide if a full wash or bottom wash is required, and if the washing is to be done by automated washing or by a handheld hose. Decide if the washing operation is to be carried out at berth, within or outside the port limits taking account of any restrictions on the discharge of cargo residues, as well as anticipated weather and sea state conditions. Always obtain written permission from the port authority to discharge hold wash water if the intention is to conduct hold washing within the port or 12 nautical mile (13.8 mi; 22 km) limits.

When washing the cargo holds, start by washing the hatch covers, hold top, bottom, and sides. Scrub the hatch cover compression bars and rubbers, if necessary, to remove any cargo traces or residues. Wash down the hatch coamings and hold deck heads. Wash the hold sides, paying particular attention to the hopper angles, pipe guards, brackets, and other non-vertical surfaces. Scrub locally and/or rewash to remove stubborn dirt, grease, and oils. Wash the deck plate and scrape up any loose rust scale. Thoroughly flush out the bilges. Rinse the holds with freshwater when possible; this helps reduce corrosion and facilitates the stowage of cargoes which cannot come into contact with salt (such as metal ores). Dry the holds by ventilating, by opening the holds, and by mopping up puddles as necessary. When washing is not necessary, sweep the holds instead. Scrape and sweep the holds and lift any residues from the hold when washing is not possible. Clean and disinfect the bilge wells. Flush out the sounding pipes and thermometer pipes. Test the bilge suctions if not used for washing. Test the bilge non-return valves, where fitted. Test the bilge high level alarms, where fitted. Inspect the holds for cleanliness, insect infestation, rodents, leaks, and structural damage. Remedy any defects where necessary. Wrap the bilge cover plates with burlap and seal with tape or cement. Reseal any manhole covers which have been opened or disturbed. Isolate the hold lighting, and lighting in compartments connected to the holds, when this precaution is required for intended cargo. Lime wash the bulkheads and tank top if required. If the holds are painted, or touched up, before a cargo of foodstuffs is carried, ensure that a paint compliance certificate can be produced for the paint used. In ballast holds, close and secure the cover plates for ballast suctions, and open the bilge suctions and CO_2 injection lines.

With respect to the disposal of residues from previous cargoes, attention is drawn to the MARPOL Convention which governs the discharge of pollution from ships at sea. In particular, Annex V of MARPOL 73/78 sets out the regulations for the prevention of pollution by garbage from ships. The disposal of plastic waste at sea is totally prohibited, while the disposal of other types of garbage is permitted only when the ship is a specified distance from land. 'Garbage' in this context is defined, by the Convention, as any form of operational waste from ships, including cargo residues from loading excess, unloading residues, and spillage. This means that under the terms of MARPOL 73/78, the discharge of cargo residues through deck and hold washing cannot occur less than 3 nautical miles (3.45 mi; 5.5 km) from the nearest shoreline. While many dry bulk cargoes may be considered harmless to the marine environment, a chief concern is the potential impact on ocean sediments and bottom-dwelling inhabitants of a build-up of materials, especially in ports and relatively shallow shipping lanes. The message is clear that the discharge of cargo residues, except in limited safety circumstances, is prohibited until the ship is more than 3 nautical miles (3.45 mi; 5.5 km) from the nearest shoreline. Indeed, from 1 July 1998, all ships of 400 gross tonnes and over are required to have an approved Garbage Management Plan and Garbage Record Book. The minimisation of cargo residue wash down and discharge should form part of the ship's Garbage Management Plan and all residue discharges should be recorded. Remember, Port State Control officers have the authority to check these records. The IMO also recommends that wherever possible cargo residues should be completely cleaned up prior to sailing and either delivered to the intended cargo space or to a recognised port reception facility. Shipboard areas where spillage is most common should be protected to enable residues to be easily recovered.

To provide some context, the Great Barrier Reef is of special environmental significance to Australians, and has been declared a World Heritage Area and Particularly Sensitive Sea Area. The MARPOL Convention has designated the Great Barrier Reef as an area in which no discharges of pollutants are permitted. This area is between the Queensland coastline and 'nearest land' defined as a line drawn between coordinates on the outer edge of the Reef. Specific distances are then measured seaward of that line. This means that ships must be at least 3 nautical miles (3.45 mi; 5.5 km) outside the Great Barrier Reef before undertaking any discharges. The MARPOL Convention should be consulted to determine the exact position of lines defining 'nearest land' in this area (refer to regulation V/1(2)).

6.4.1 Exceptions

It is understood that dry cargo residues are washed down by ships' crews for not only operational reasons but safety reasons as well. The safety of the ship, its crew, and others involved in working the ship is of vital importance. Therefore, the MARPOL Convention provides exceptions from the discharge restrictions where there is a threat to the safety of the ship and to those onboard. Referring again to our example of the Great Barrier Reef, in accordance with MARPOL regulation V/6(a), AMSA and the Great Barrier Reef Marine Park Authority will accept the cleaning of cargo residues from a vessel within the 3 nautical mile limit in any one of the following circumstances:

(1) To ensure the safe operation of a helicopter (for taking onboard a pilot or other such purpose). This exception applies only to the helicopter landing area and its immediate vicinity to avoid dust being raised by the down draft of the helicopter rotors and does not extend to the systematic wash down of the entire vessel. For additional information on helicopter operations, a copy of the *Australian Code of Practice for Ship-Helicopter Transfers* may be obtained from any AMSA office.
(2) Where there is a need to avoid navigational hazards like dust being blown onto areas such as the wheelhouse or bridge wings.
(3) Where residues cause a serious safety hazard to personnel if pillages are not cleaned from deck areas, adjacent walkways, and working areas.

There are substantial penalties for breaches of the MARPOL restrictions on discharges from ships. These can include substantial fines for the shipowner or operator, the master and crew, and where criminal activity is found, even imprisonment. Additional information on the proper disposal of dry bulk cargo residues and associated wastes can be found in chapter 5 of the volume *Bulk Carrier Practice*, published by the Nautical Institute, and in *Bulk Carriers: Guidance and Information to Shipowners and Operators*, published by the International Association of Classification Societies (IACS). Moreover, the booklet *Guidelines for the Preparation of Garbage Management Plans* produced by the International Chamber of Shipping (ICS) also provides useful guidance. For further information on Australian regulations contact

The General Manager
Maritime Safety and Environment Strategy
Australian Maritime Safety Authority
GPO Box 2181
Canberra ACT 2601
Australia

6.5 SUMMARY

The carriage of solid bulk cargo comprises several risks, which can be quite costly at times. Several bulk carriers have disappeared with lives lost after loading solid bulk cargo. Officers must pay attention to the loading process, the risks of structural damage to the vessel, and the hazards associated with the cargo carried. Though damage to the cargo is less likely, damage to the ship and personnel are always a possibility that cannot be ignored at any cost.

Chapter 7

Principles for the Safe Handling, Stowage, and Carriage of Liquid Bulk Cargo

7.1 INTRODUCTION

Liquid cargoes in bulk are generally divided into one of two categories:

(1) Vegetable and edible liquids
(2) Petroleum products

Vegetable and edible liquids are different from the mineral and crude oils that are transported by oil tankers. Vegetable oils usually solidify at room temperature or below, so heating coils are required in the tanks containing such cargoes. The carriage of edible oils involves a strict standard of cleanliness. Appropriate certificates as to the seaworthiness and cleanliness of the vessel are necessary before loading can commence. Every effort should be made to avoid the ingress of foreign matter into the storage tanks. Contamination is one of the greatest risks associated with the carriage of vegetable oils. After cleaning of the tank, prior to loading, the tank will be steamed out for not less than 24 hours and then washed under pressure. The watertightness of the tank must then be checked, usually by filling it up with water. Sometimes caustic soda and sand is used to clean the tank. Heating coils are normally fitted both at the bottom and at the top of the tank to allow for a speedy discharge. The carrying temperatures of vegetable oils differ somewhat with the class of oil. It is usual for shippers to specify the carrying temperature. Discharge is done either by shore plants in the form of independent steam pumps operating at the tank top flat or by a combined suction and discharge pump, which may be lowered into the tank. Some general cargo ships may have separate edible oil tanks, though today tanktainers – an oil tank enclosed in a container frame – are now common (Figure 7.1).

Alternatively, petroleum liquids are divided into one of two classes: 'dirty' and 'clean' oils. Dirty oils include mineral products such as crude oil, fuel oil, creosote, and bitumen, whereas clean products include benzine, naphtha, and aviation spirit. Due to the nature of these products, the risks involved in

DOI: 10.1201/9781003354338-8

Figure 7.1 Typical tanktainer.
Source: Author's own.

the carriage of petroleum products is exceptionally high. The three primary hazards are as follows:

(1) Risk of fire
(2) Risk of explosion
(3) Risk of toxic inhalation

Moreover, the carriage of any petroleum product also carries the risk of pollution of the environment. There is no doubt that the carriage of petroleum products covers a whole array of risks that the cargo officer must always remain cognizant of when involved with any cargo-carrying operation. When petroleum is ignited, it is not the liquid itself that ignites but rather the vapours that are present just above the surface of the liquid. It is these vapours which burn, and not the liquid. The volatility of the liquid denotes the amount of vapour that is given off. The volatility of the cargo depends largely on ambient temperature. When mixed with the right amount of oxygen, in the presence of an ignition source, ignition will occur. If there is too much vapour present or insufficient vapour is present, the mixture is said to be too rich or too lean. In between these two points is the sweet spot where the vapours will freely ignite. This is called the flammable limit. The expansion of the gases when burning can be the cause for explosion if enclosed within compartments. These fumes can also be extremely toxic when inhaled. Caution should therefore be exercised as some of these gases deaden the sense of smell, thus removing any signs of danger to the human being.

Without getting to embroiled in the science of fire, it is worth spending a few moments discussing some of the key concepts that cargo officers need to understand to work safely. The first term is *auto-ignition*. This is the lowest temperature to which a solid, gas, or liquid must be raised to cause a self-sustained combustion without an external source of ignition, such as a spark or flame. The *flammable range*, as we mentioned earlier, is the range of hydrocarbon gas concentrations in air between which the vapours can and cannot ignite. If the concentration of vapours is too low, the concentration is lean; if the vapours are too high, the concentration is too rich. Within these outer limits, the vapours will ignite. The *flashpoint* is the lowest temperature at which a liquid gives off sufficient vapours to form a flammable gas mixture near the surface of the liquid. *Gas free* refers to a tank or compartment in which sufficient fresh air has been introduced to lower the level of any flammable, toxic, or inert gas to that required for a specific purpose, for example, hot work, safe entry, etc. *Inert gas* is a type of gas or mixture which is incapable of supporting combustion. This includes nitrogen, CO_2, and flue gas. A *combustible gas indicator* is an instrument for measuring the composition of hydrocarbon vapours and/or vapour mixtures, usually giving the result as a percentage of the lower flammable limit. The *lower flammable limit* is the concentration of a hydrocarbon vapour in air below which there is insufficient hydrocarbons to support and propagate combustion. The opposite to this is the *upper flammable limit*, in which the concentration of a hydrogen gas in air above which there is insufficient air to support and propagate combustion. *Static electricity* is the electricity produced on dissimilar materials through physical contact and separation. The *ignition point* is the lowest temperature at which it must be raised so that vapour is evolved at that rate which permits continuous burning when a flame is momentarily applied.

Like combustion, there are specific terms related to flammability and vapour pressure that cargo officers must be aware of. These include *non-volatile cargoes* which have a flashpoint above 60°C (140°F) and thus are handled at temperatures below this. These are normally safe products. Alternatively, v*olatile cargoes of low vapour pressure* are those cargoes whose flashpoint is lower than 60°C (140°F), but usually above handling temperature. *Volatile cargoes of intermediate vapour pressure* have a flashpoint which is lower than 60°C (140°F) and usually lower than the handling temperature. Thus, large quantities of vapours are given off. The relatively low vapour pressure makes it comparatively safe to handle. *Volatile cargoes of high vapour pressure* are usually volatile enough to produce sufficient vapour to keep the space well above the flammable range, that is, the air is too rich, but when loaded into an empty space and when venting, flammable mixtures result.

7.2 CARRIAGE OF PETROLEUM PRODUCTS

Specially designed ships are used to carry petroleum products in bulk. In designing these tankers, special consideration must be given to factors such as the inherent risks of fire, the rate of loading and discharging, the installation of cofferdams at the ends of cargo tanks, ventilating pipes and ullage sounding pipes, presence of heating coils, tank cleaning equipment and procedures, presence of dedicated ballast tanks, and cargo cross-contamination. During the loading operations, apart from the risks of fire and explosion, possible contamination of the cargo and accumulation of static charges make the operation especially hazardous. It should be obvious that clean oils must never be contaminated by being loaded into the same tanks that were previously used to carry dirty oils and which have not been thoroughly cleaned. In 1992, the IMO amended the MARPOL Convention to make it mandatory for new tankers (i.e. those ordered after 6 July 1993) of 5,000 tonnes or more to be double hulled. A revised phase-out schedule for single hulled tankers was adopted in April 2001 with the intention of phasing out of all single hulled tankers by 2015. For this reason, various pipeline configurations exist. These serve specific purposes and eliminate the risks of cargo contamination. The *direct system* consists of three bottom lines and pumps, with each one serving one part of the vessel. Contamination is less likely as it is easier to isolate each section. The disadvantage with this system, however, is that it limits the number of types of products that can be carried to a maximum of three. The *ring main system* consists of a single pipeline which runs through the wing tanks, which are linked to each tank by a crossover valve at each end of the vessel and split at the pump rooms. Contamination is more likely in this system where the valve configuration is not set properly. A greater number of small parcels is possible with this set up. The *free flow system* consists of a pipeline which feeds round the ship with sluice valves at the bottom of the inter-tank bulkheads. These sluice valves permit the flow of oil to or from each tank directly, as required. The main advantage of this system is the very high discharge rates that are achievable. On the other hand, the potential for tank overflowing is greater with this configuration when compared to the direct and ring main systems.

Some of the precautionary measures to be taken during loading include ensuring the tanker manifolds are connected to the shore installation by hoses. These hoses are fitted with an insulating section, which prevents the flow of stray currents caused by the anodic protection of the vessel. These currents, if in contact with any spark, will result in major explosions. The build-up of static electricity in the hoses from the flow of liquid is also another serious cause for concern. The lines are normally bonded to prevent this from occurring. Whereas a high loading rate is desirable, extreme care must be taken to avoid spillage and to this end efficient communication must be maintained between the ship's crew and the shore

Figure 7.2 Reach Stacker lifting a tanktainer.

staff. Loading should commence at a slow rate and be seen to be proceeding satisfactorily before the rate is increased. Adequate notice should
be given to the shore staff prior to topping up and completion. Liquids
should never be dropped from the top. When a tank has been closed, the
ullage of that tank should be checked to ensure that the valve is operating properly. Another source of danger is the high rate of loading that
affects the structural soundness of the vessel. Over-stressing the vessel
can happen easily and this should be closely monitored. With volatile
cargoes, it must always be borne in mind that gaseous vapour surrounds
any opening, thus encouraging an explosive risk. The same is true with
tank cleaning or ballasting after discharging such cargoes. Broadly, the
loading and discharging of multi-cargoes is a matter of 'valve familiarity
and crossovers'. It is unwise to load volatile oils directly through the tank
hatches. Only non-volatile oils should be so loaded and then only if the
tanks are certified as gas-free (Figure 7.2).

7.3 CARGO-HANDLING PROCEDURES FOR CLEAN OILS

During loading of clean oils, the flow rate should be less than 1 m/s (3.2 ft/s)
in the pipe until the level of oil rises above the line inlet (Figure 7.3). At this
point, all splashing and turbulence should cease. Where inert gas is used,

Figure 7.3 IBC tanks.

the risk of sparks exists at the tank openings unless sampling and ullaging gear is earthed directly to the hull. Most of these oils have a high vapour pressure, therefore avoid loading in still weather, and utilise low initial and topping off rates. Avoid venting off at deck level wherever possible. If the oil is hot, avoid loading or ensure that venting is provided well above the superstructure. During discharging, pay particular attention to the vessel's moorings. When the shore installation is ready to receive the cargo, open the suction and gate valves and start the pump. Watch the back pressure in case there is any obstruction in the shorelines.

7.4 CARGO-HANDLING PROCEDURES FOR DIRTY OILS

In summary, the cargo-handling procedures require a checklist to be completed to ensure preliminary inspections have been completed accordingly. The loading and/or discharging plan must be drawn up and agreed, considering the following:

(1) Changeover between tanks
(2) Avoidance of contamination between grades

(3) Avoidance of pollution by spillage
(4) Clearance of pipelines
(5) Stress and trim during cargo transfer
(6) Loading and/or discharging rates
(7) Weather conditions leading to interruption of work
(8) Allowance for crude oil washing (COW) if carried out

A formal agreement that the ship and terminal are ready to commence work. Checks that after the pumping operation has commenced, the cargo is entering right tank. Regular checks of the hose and pipeline pressures. Ullage is carefully monitored and managed to avoid gas build-ups. Commencement of loading at a slow rate until the correct cargo flow is achieved and verified. Topping off. As each tank nears the final ullage, the valve to the next tank is cracked open increasingly until the first tank is full. Checking the ullages from time to time to verify that the valves to the full tanks have been properly closed. Precautions to ensure static accumulation hazards are monitored until the surface of the oil has settled and the electric charge has dispersed. No dipping or sampling with metal tape or cans is permitted until a minimum of 30 minutes has elapsed after the loading operation has ceased. Ensure that the cargo loading and/or discharge plan is followed scrupulously to avoid over-stressing and contamination.

7.5 CRUDE OIL WASHING

The carriage of petroleum products comes with two exclusive concepts: tank cleaning (or, more precisely, crude oil washing) and inerting (that is, to make the atmosphere in the tank unable to sustain combustion). Crude oil washing (COW) is a process whereby some of the cargo is used to clean a tank, which is actually being discharged. Fixed tank cleaning equipment is used to remove the waxy asphaltic deposits, which the cargo has left on the tanks sides. It has proved to be more effective than water as the oils act to disperse and suspend the sediments in the oil, which tends to restore the cargo to its original condition. To carry out crude oil washing, three requirements are needed:

(1) The vessel must have an efficient inert gas system.
(2) A fixed tank cleaning installation is necessary.
(3) There must be an effective monitoring system.

Before crude oil washing is commenced, any water in the cargo tank should be drawn out. The operation should only be carried out when the tank's oxygen level is below 8%. With the multistage washing method, the tank sides are washed as the cargo level falls and the bottom is washed as the tank empties. The machines are provided with the oil via junction lines from the discharge lines of the main cargo pumps. Some of the advantages of crude oil washing are reduced risks of pollution, reduced tank cleaning times, the removal of sludge, less salt water sent to the refinery, reductions in the corrosion process, increases in cargo discharges, increased cargo-carrying capacity, increased time available for maintenance, and increased discharge rates. Alternatively, the disadvantages of crude oil washing include additional crew training requirements, an increased workload when in port, potential structural damage caused by high-pressure jets striking the tank membranes, and possible reduced rates of discharge on some types of very large crude carriers (VLCCs).

7.6 PRINCIPLES OF THE INERT GAS SYSTEMS

There are four primary factors which contribute to cargo tank explosions:

(1) Hydrocarbon vapours
(2) Oxygen in the right proportion
(3) Sources of ignition
(4) Chemical reaction

Withdrawing any one of the four constituents will reduce or eliminate the risk of a cargo tank explosion. The introduction of inert gas into the cargo tank reduces the oxygen content to a lower level, thereby reducing the hydrocarbon vapour concentration in the atmosphere to a safe proportion. Thus, two factors have been made innocuous, and protection against a tank explosion has been achieved. There are several benefits to using inert gas in cargo operations onboard tankers. Apart from the obvious safety reason, the use of inert gas allows for an increase in discharging rates, allows for the carriage of certain types of cargo that react with oxygen, and helps in corrosion control by displacing oxygen, which is a major cause of corrosion.

7.7 PREPARATION OF CARGO SPACES FOR THE RECEPTION OF BULK LIQUID CARGO

When preparing to receive bulk liquid cargoes such as crude oil, it is important to ensure the space is fully inerted (i.e. by replacing the atmosphere in the tank with a gas that does not have sufficient oxygen to support combustion). The cargo space must be washed, usually using a tank washing machine. The cargo space must then be gas-freed, sampled, and certified as gas-free (if necessary, carrying out forced ventilating of the compartment). Usually, an independent surveyor or chemist is required to issue the gas-free certificate. The cargo space should be inspected for signs of defects and/or damage to suctions, submerged valves and pumps, cathodic protection, and the tank coatings. All fixed fire-extinguishing installations must be checked and confirmed to be in good working condition. All pressure valves, spark arrestors, and gauges must also be checked. All seals on tank lids, ullage ports, and tank washing machine openings must be checked and renewed, where necessary. Before loading, the cargo space is to be inerted again if it was gas-freed to prevent any risk of developing a flammable mixture. The following guidance is adapted from the *Code of Safe Working Practices for Australian Seafarers* (by kind permission of AMSA).

In relation to oil tankers and product carriers, particular attention is drawn to the importance of the *International Safety Guide for Oil Tankers and Terminals (ISGOTT)* which provides comprehensive information on the safe operation of tankers. Shipowners should provide seafarers employed on tankers with appropriate training and instructions in the relevant operational and safety requirements associated with their duties and emergency situations. For each operation the master should designate a competent officer who is familiar with the safe operation of tankers. The master should ensure that the designated officer has available an adequate number of competent persons. Particular attention is drawn to the following specific issues:

(1) The need for a well-structured onboard safety policy backed up by the appropriate safety committee with designated responsibilities (refer to chapter 2 of the *Code of Safe Working Practices for Australian Seafarers* for more information).

(2) The need for strict smoking and hot work policies.

(3) The need for crew members to fully understand the hazardous nature of cargoes carried.

(4) The need for crew members to be aware of the inherent dangers of cargo pump rooms. Pump rooms, by virtue of their location, design, and operation, constitute a particular hazard and therefore necessitate special precautions.

(5) The need for crew members to be made aware of the carcinogenic health hazards resulting from exposure to minor concentrations of benzene vapour in the air. This hazard can result from breathing vapours of benzene containing cargoes such as gasoline, JP-4, and some crude oils.

(6) The need to ensure that seafarers are made aware of the safety precautions and emergency action to be taken in the event of spillage.

With specific regard to bulk chemical tankers, the points listed earlier may apply equally, in addition to those listed here. Ships intended for the carriage of chemicals should carry only those chemicals for which their construction and equipment are suitable, and which are specified on the certificate of fitness. Particular attention is drawn to the importance of having comprehensive information on the safe operation of chemical tankers. Only approved documentation should be used. MSDS should be provided and be freely available for all chemical cargoes carried. Shipowners should provide seafarers employed on chemical tankers with specialised training and instructions in the safe carriage of all chemicals which the ship may be required to carry and the relevant operational and safety requirements associated with their duties and emergency situations. For each operation, the master should designate a competent officer who is familiar with the safe operation of chemical tankers. The master should ensure that the designated officer has available an adequate number of experienced seafarers. Particular attention is drawn to the following needs:

(1) Ensure that any cargo offered is listed in the shipping documents by the correct technical name.
(2) Ensure that where a cargo is a mixture, an analysis is provided indicating the dangerous components which contribute significantly to the hazard of the product. This information should be available onboard, and freely accessible to all concerned.
(3) Ensure that a full description of a cargo's physical and chemical properties is supplied with each cargo loaded.
(4) Ensure that seafarers are made aware of the safety precautions and emergency action to be taken in the event of spillage or crew exposure to possible contamination by chemicals.
(5) Ensure that cargoes requiring stabilisers or inhibitors, and which are not accompanied by the required certificates, are not accepted for shipment.
(6) Carry out emergency drills using protective equipment and safety and rescue devices at regular intervals.
(7) Plan effective first aid treatment in the event of accidental personal contact.

For each operation, the vessel master should designate a competent officer who is familiar with the safe operation of tankers. In addition, the master should ensure that the designated officer has available an adequate number of suitably trained and experienced seafarers.

In this chapter, we have discussed some of the main points around the carriage of liquid bulk cargoes. It is important to recognise there are very separate methods for handling vegetable and edible liquids and petroleum-type products. It is also worth noting that within the second category (petroleum products), there are two further subcategories – those being dirty and clean oils. It is crucial that every effort is taken to ensure different categories and classes of cargoes are kept separated to avoid contamination, and that all necessary safety precautions are taken to reduce and eliminate the risk of ignition and explosion. In the third and final chapter on the safe handling, stowage, and carriage of bulk cargoes, we will examine the carriage of liquefied gases in bulk.

Chapter 8

Principles for the Safe Handling, Stowage, and Carriage of Liquefied Gases

8.1 INTRODUCTION

It is common during the offshore oil extraction process to produce natural gases as a by-product. These gases consist of natural gases such as methane and petroleum gases such as propane and butane. These gases are extracted and shipped onshore. Before the gases are shipped, they are compressed under high pressure into liquid form. Petroleum gases (propane and butane) are then classified as LPG and natural gas (methane) is classified as LNG. Because of their chemical properties (usually a very low boiling point), these cargoes need special care to be transported safely. Much use is made of the laws of physics to establish the best possible method of transportation, as pressure, temperature, and volume are all related to each other. Thus, by pressurising or cooling the gases down (or through a combination of both), the natural gases can be transported onboard specially designed vessels. Some of the key definitions used in the carriage of liquefied gas cargoes are discussed here:

- *Critical temperature:* This is the temperature above which the gas cannot be liquefied by pressure. For methane, this is 82°C (179.6°F) and for propane, 97°C (206.6°F).
- *Vapour pressure:* This is the pressure exerted by the vapour of the liquid. It is proportional to temperature. For example, at the critical temperature, methane requires about 4.6 MPa to be liquefied.
- *Boiling point:* The boiling point is the temperature at which the substance will cease to be a liquid and starts giving off-vapours. For example, methane's boiling point is –161°C (–257.8°F), propane is –42°C (–42°F) and butane is 0°C (32°F).
- *Fully pressurised:* For LPG and ammonia, which is also classified as a petroleum gas, tanks are made from carbon steel, which is suitable for minimum service temperatures of –5°C (23°F). There is no refrigeration plant on these types of vessels which have a maximum carrying capacity of 2,000 m³.

DOI: 10.1201/9781003354338-9

- *Semi-pressurised and partially refrigerated:* These vessels have a maximum working pressure of 0.8 MPa. The service temperature is around –5°C (23°F). The wall thickness is reduced because of the refrigeration plant, which uses the cargo as the refrigeration medium to maintain the vapour pressure.
- *Semi-pressurised and fully refrigerated:* These vessels have a minimum service temperature of –45°C (–49°F). The tank surface is manufactured from low-temperature carbon or nickel alloy steel and has a working pressure of 0.5–0.8 MPa. The tanks are normally cylindrical or spherical.
- *Fully refrigerated:* These vessels are subdivided into three groups:

(1) Fully refrigerated LPG tankers
(2) Fully refrigerated ethylene carriers
(3) LNG carriers

The minimum service temperature of LPG is around –50°C (–58°F), whilst for LNG it is around –164°C (–263.2°F). The various tank types cater for the need for insulation and the enormous pressure that the cargo exerts on the tank structure.

8.2 CARGO OPERATIONS ON GAS CARRIERS

Gas carriers require very specific phases to be taken when preparing to receive their cargo. These are briefly outlined here. The first phase is the *drying* step. The presence of water in any part of the cargo-handling system will prevent correct operation and may be difficult to remove after cooling down. Moisture traps must be avoided. Drain and purge cocks must be provided at both high and low points. The dew point of any air or inert gas must be lower than the minimum cargo service temperature to avoid condensation from building up inside the tank. The second phase involves *inerting* the tank. This is done to reduce the oxygen content in the cargo-handling system to prevent flammable mixtures and to prevent a chemical change (for example, with ethylene). Pumps, compressors, cargo instrument lines, and cargo tanks and pipelines must be inerted. In LNG designs, the space around the tanks must also be inert. The third phase involves *purging* the tanks. Inert gases are insoluble in hydrocarbon and ammonia cargoes and thus will not dissolve. They tend to block lines and machinery and thus must be purged out by cargo vapour. The fourth phase is *pre-cooling*. This is done to reduce stress in the cargo tank structures caused by local differences in temperature. It also reduces the quantity of vapour released during loading as the cold liquid cargo level rises along the tank walls. Once these initial phases are complete, the loading of the cargo can begin.

When *loading* the cargo, the rate is governed by the rate at which the cargo vapours can be safely disposed of. Some loading jetties are equipped with dedicated vapour return facilities. During loading, the cargo warms up in the delivery pipeline and whilst entering the tank. Some vapourises and reduces the temperature of the rest of the cargo. This means it is necessary to *condition the cargo for the voyage*. The warmer, less dense liquid rises and maintains a certain vapour pressure in the tank. Operation of the reliquefication plant and withdrawal of the vapour to lower tank pressure causes surface boiling and self-refrigeration at the liquid surface. Some vessels use this 'boil off' to supplement the fuel used in the ship's main engine.

When it comes to *discharging the cargo*, the liquefied gases at their boiling temperature are difficult to pump as they have poor lubrication and cooling properties, and easily boil. This can cause pump cavitation and equipment damage. Therefore, they must be kept cold. If the same cargo is to be loaded again, then the tanks must be maintained in a condition compatible with that cargo. This is achieved by retaining a small quantity of the product onboard. If the ship is to carry a different cargo, then it must be gas freed. This process is referred to as *conditioning on ballast passages*. *Gas-freeing* involves removing all liquids, warming the tank, and inerting the tank and cargo system. Air is then introduced to the tank (Figure 8.1).

Figure 8.1 Interior of a typical LNG membrane tank.

Source: Author's own.

The following guidance for liquefied natural and petroleum gas carriers is adapted from the *Code of Safe Working Practices for Australian Seafarers* and is reproduced courtesy of the Australian Maritime Safety Authority:

Ships intended for the carriage of liquefied gas should carry only those liquids for which its construction and equipment are suitable, and which are specified on the certificate of fitness. Particular attention is drawn to the importance of the *ICS Publication Tanker Safety Guide (Liquefied Gas)* and the book *Liquefied Gas Handling Principles on Ships and in Terminals*, which provides comprehensive information on the safe operation of liquefied gas carriers. Shipowners should provide seafarers employed on liquefied gas carriers with appropriate raining and instructions in the relevant operational and safety requirements associated with their duties and emergency situations. Comprehensive operating instructions should be provided concerning the particular ship and cargo. For each operation, the master should designate a competent officer who is familiar with the safe operation of liquefied gas carriers. The master should ensure that the designated officer has available an adequate number of experienced seafarers. Particular attention is drawn to the following needs:

(1) Ensure that a full description of the cargo's physical and chemical properties is supplied with each cargo loaded.
(2) Ensure that seafarers are made aware of the safety precautions and emergency action to be taken in the event of spillage.
(3) Plan effective first aid treatment due to possible physical contact with liquefied gases or cold cryogenic pipelines, some of which can be at a temperature of –160°C (–256°F).
(4) Carry out emergency drills at regular intervals using personal protective equipment and safety and rescue devices.

8.3 GUIDELINES FOR COMPLETING THE SHIP TO SHORE SAFETY CHECKLIST

(15) Specific guidance exists for the completion of ship to shore liquid bulk cargo transfers. This guidance contains several key questions that should be answered as part of the vessel's pre-loading and pre-discharging operations1) Is the ship securely moored?

In answering this question, due regard should be given to the need for adequate rendering arrangements. Ships should remain adequately secured in their moorings. Alongside piers or quays, ranging of the ship should be prevented by keeping all mooring lines taut; attention should be given to the movement of the ship caused by wind, currents, tides, or passing ships and the operation in progress. The wind velocity at which loading arms should be disconnected, cargo operations stopped, or the vessel berthed should be stated. Wire ropes and fibre ropes should not be used together in the same direction (i.e. breasts, springs, head or stern) because of the difference in their elastic properties. Once moored, ships fitted with automatic tension winches should not use such winches in the automatic mode. Means should be provided to enable quick and safe release of the ship in case of an emergency. In ports where anchors are required to be used, special consideration should be given to this matter. Irrespective of the mooring method used, the emergency release operation should be agreed, taking into account the possible risks involved. Anchors not in use should be properly secured.

(2) Are emergency towing wires correctly positioned?

Emergency towing wires (fire wires) should be positioned on both the offshore bow and quarter of the ship. At a buoy mooring, emergency towing wires should be positioned on the side opposite to the hose string. There are various methods for rigging emergency towing wires currently in use. Some terminals may require a particular method to be used and the ship should be advised accordingly.

(3) Is there safe access between ship and shore?

The access should be positioned as far away from the manifolds as practicable. The means of access to the ship should be safe and may consist of an appropriate gangway or accommodation ladder with a properly secured safety net fitted to it. Particular attention to safe access should be given where the difference in level between the point of access on the vessel and the jetty or quay is large or likely to become large. When terminal access facilities are not available and a ship's gangway is used, there should be an adequate landing area on the berth so as to provide the gangway with a sufficient clear run of space and so maintain safe and convenient access to the ship at all states of tide and changes in the ship's freeboard. Near the access ashore, appropriate life-saving equipment should be provided by the terminal. A lifebuoy should be available onboard the ship near the gangway or accommodation ladder. The access should be safely and properly illuminated during darkness. Persons who have no legitimate business onboard, or who do not have the master's permission, should be refused access to the ship. The terminal should control access to the jetty or berth in agreement with the ship.

(4) Is the ship ready to move under its own power?

The ship should be able to move under its own power at short notice, unless permission to immobilise the ship has been granted by the Port Authority and the terminal manager. Certain conditions may have to be met for permission to be granted.

(5) Is there an effective deck watch in attendance onboard and adequate supervision on the terminal and on the ship?

The operation should be under constant control on both ship and shore. Supervision should be aimed at preventing the development of hazardous situations; if however such a situation arises, the controlling personnel should have adequate means available to take corrective action. The controlling personnel on ship and shore should maintain an effective communication with their respective supervisors. All personnel connected with the operations should be familiar with the dangers of the substances handled.

(6) Is the agreed ship/shore communication system operative?

Communication should be maintained in the most efficient way between the responsible officer on duty on the ship and the responsible person ashore. When telephones are used, the telephone both onboard and ashore should be continuously manned by a person who can immediately contact his respective supervisor. Additionally, the supervisor should have a facility to override all calls. When VHF systems are used, the units should preferably be portable and carried by the supervisor or a person who can get in touch with his respective supervisor immediately. Where fixed systems are used, the guidelines for telephones should apply. The selected system of communication, together with the necessary information on telephone numbers and/or channels to be used, should be recorded on the appropriate form. This form should be signed by both ship and shore representatives. The telephone and portable VHF systems should comply with the appropriate safety requirements.

(7) Has the emergency signal to be used by the ship and shore been explained and understood?

The agreed signal to be used in the event of an emergency arising ashore or onboard should be clearly understood by shore and ship personnel.

(8) Have the procedures for cargo, bunker, and ballast handling been agreed?

The procedures for the intended operation should be pre-planned. They should be discussed and agreed upon by the ship and shore representatives prior to the start of the operations. Agreed arrangements should be formally recorded and signed by both ship and terminal representatives. Any change in the agreed procedure that could affect the operation should be discussed by both parties and agreed upon. After agreement has been reached by both parties, substantial changes should be laid down in writing as soon as possible and in sufficient time before the change in procedure takes place. In any case, the change should be laid down in writing within the working period of those supervisors onboard and ashore in whose working period agreement on the change was reached. The operations should be suspended and all deck and vent openings closed on the approach of an electrical storm. The properties of the substances handled, the equipment of ship and shore installation, the ability of the ship's crew and shore personnel to execute the necessary operations and to sufficiently control the operations are factors which should be taken into account when ascertaining the possibility of handling a number of substances concurrently. The manifold areas both onboard and ashore should be safely and properly illuminated during darkness. The initial and maximum loading rates, topping off rates, and normal stopping times should be agreed, having regard to the following:

(a) The nature of the cargo to be handled.
(b) The arrangement and capacity of the ship's cargo lines and gas venting systems.
(c) The maximum allowable pressure and flow rate in the ship/shore hoses and loading arms.
(d) Precautions to avoid accumulation of static electricity.
(e) Any other flow control limitations.
(f) A record to this effect should be formally made as above.

(9) Have the hazards associated with toxic substances in the cargo being handled been identified and understood?

Many tanker cargoes contain components which are known to be hazardous to human health. In order to minimise the impact on personnel, information on cargo constituents should be available during the cargo transfer to enable the adoption of proper precautions. In addition, some port states require such information to be readily available during cargo transfer and in the event of an accidental spill. The information provided should identify the constituents by chemical name, name in common usage, UN number, and the maximum concentration expressed as a percentage by volume.

(10) Has the emergency shutdown procedure been agreed upon?

An emergency shutdown procedure should be agreed upon between ship and shore, formally recorded and signed by both the ship and terminal representative. The agreement should state the circumstances in which operations have to be stopped immediately Due regard should be given to the possible introduction of dangers associated with the emergency shutdown procedure.

(11) Are fire hoses and firefighting equipment onboard and ashore positioned and ready for immediate use?

Firefighting equipment both onboard and ashore should be correctly positioned and ready for immediate use. Adequate units of fixed or portable equipment should be stationed to cover the ship's cargo deck and on the jetty. The ship and shore fire main systems should be pressurised or be capable of being pressurised at short notice. Both ship and shore should ensure that their fire main systems can be inter-connected in a quick and easy way utilising, if necessary, the international shore fire connection.

(12) Are cargo and bunker hoses/arms in good condition, properly rigged, and appropriate for the service intended?

Hoses should be in a good condition and properly fitted and rigged to prevent strain and stress beyond design limitations. All flange connections should be fully bolted, and any other types of connections should be properly secured. It should be ensured that the hoses/arms are constructed of a material suitable for the substance to be handled accounting for its temperature and the maximum operating pressure. Cargo hoses should be properly marked and identifiable with regard to their suitability for the intended operation.

(13) Are scuppers effectively plugged and drip trays in position, both onboard and ashore?

Where applicable, all scuppers onboard and drain holes ashore should be properly plugged during the operations. Accumulation of water should be drained off periodically. Both ship and jetty manifolds should ideally be provided with fixed drip trays; in their absence, portable drip trays should be used. All drip trays should be emptied in an appropriate manner whenever necessary but always after completion of the specific operation. When only corrosive liquids or refrigerated gases are being handled, the scuppers may be kept open, provided that an ample supply of water is available at all times in the vicinity of the manifolds.

(14) Are unused cargo and bunker connections proper secured with blank flanges fully bolted?

Unused cargo and bunker line connections should be closed and blanked. Blank flanges should be fully bolted and other types of fittings, if used, properly secured.

(15) Are sea and overboard discharge valves, when not in use, closed and visibly secured?

Experience shows the importance of this item in pollution avoidance on ships where cargo lines and ballast systems are interconnected. Remote operating controls for such valves should be identified in order to avoid inadvertent opening. If appropriate, the security of the valves in question should be checked visually.

(16) Are all cargo and bunker tank lids closed?

Part from the openings in use for tank venting (refer to question 17) all openings to cargo tanks should be closed and gas tight. Except on gas tankers, ullaging and sampling points may be opened for the short periods necessary for ullaging and sampling. Closed ullaging and sampling systems should he used where required by international, national, or local regulations and agreements.

(17) Is the agreed tank venting system being used?

Agreement should be reached, and recorded, as to the venting system for the operation, taking into account the nature of the cargo and international, national, or local regulations and agreements. There are three basic systems for venting tanks:

(a) Open to atmosphere via open ullage ports, protected by suitable flame screens
(b) Fixed venting systems which include inert gas items
(c) To shore through other vapour collection systems

(18) Has the operation of the P/V valves and/or high-velocity vents been verified using the checklift facility, where fitted?

The operation of the P/V valves and/or high-velocity vents should be checked using the testing facility provided by the manufacturer. Furthermore, it is imperative that an adequate check is made, visually or otherwise at this time to ensure that the checklift is actually operating the valve. On occasion a seized or stiff vent has caused the checklift drive pin to shear and the ship's personnel to assume, with disastrous consequences, that the vent was operational.

(19) Are hand torches of an approved type?
And

(20) Are portable VHF/UHF transceivers of an approved type?

Battery-operated hand torches and VHF radio-telephone sets should be of a safe type which is approved by a competent authority. Ship/shore telephones should comply with the requirements for explosion-proof construction except when placed in a safe space in the accommodation. VHF radio-telephone sets may operate in the internationally agreed wave bands only. The aforementioned equipment should be well-maintained. Damaged units, even though they may be capable of operation, should not be used.

(21) Are the ship's main radio transmitter aerials earthed and radars switched off?

The ship's main radio station should not be used during the ship's stay in port, except for receiving purposes. The main transmitting aerials should be disconnected and earthed. Satellite communications equipment may be used normally unless advised otherwise. The ship's radar installation should not be used unless the master, in consultation with the terminal manager, has established the conditions under which the installation may be used safely.

(22) Are electric cables to portable electrical equipment disconnected from power?

The use of portable electrical equipment on wandering leads should be prohibited in hazardous zones during cargo operations and the equipment preferably removed from the hazardous zone. Telephone cables in use in the ship/shore communication system should preferably be routed outside the hazardous zone. Wherever this is not feasible, the cable should be so positioned and protected that no danger arises from its use.

(23) Are all external doors and ports in the accommodation closed?

External doors, windows, and portholes in the accommodation should be closed during cargo operations. These doors should be clearly marked as being required to be closed during such operations, but at no time should they be locked.

(24) Are window-type air-conditioning units disconnected?

(25) Are air-conditioning intakes which may permit the entry of cargo vapours closed?

Window-type air-conditioning units should be disconnected from their power supply. Air-conditioning and ventilator intakes which are likely to draw in air from the cargo area should be closed. Air-conditioning units which are located wholly within the accommodation, and which do not draw in air from the outside, may remain in operation.

(26) Are the requirements for the use of galley equipment and other cooking appliances being observed?

Open fire systems may be used in galleys whose construction, location, and ventilation system provides protection against entry of flammable gases. In cases where the galley does not comply with the above, open fire systems may be used provided the master, in consultation and agreement with the terminal representative, has ensured that precautions have been taken against the entry and accumulation of flammable gases. On ships with stern discharge lines which are in use, open fire systems in galley equipment should not be allowed unless the ship is constructed to permit their use in such circumstances.

(27) Are smoking regulations being observed?

Smoking onboard the ship may only take place in places specified by the master in consultation with the terminal manager or his representative. No smoking is allowed on the jetty and the adjacent area except in buildings and places specified by the terminal manager in consultation with the master. Places which are directly accessible from the outside should not be designated as places where smoking is permitted. Buildings, places, and rooms designated as areas where smoking is permitted should be clearly marked as such.

(28) Are naked light regulations being observed?

A naked light or open fire comprises the following: flame, spark formation, naked electric light, or any surface with a temperature that is equal to or higher than the minimum ignition temperature of the products handled in the operation. The use of open fire onboard the ship, and within a distance of 25 metres of the ship, should be prohibited, unless all applicable regulations have been met and agreement reached by the port authority, terminal manager, and the master. This distance may have to be extended for ships of a specialised nature such as gas tankers.

(29) Is there provision for an emergency escape?

In addition to the means of access referred to in question 3, a safe and quick emergency escape route should be available both onboard and ashore. Onboard the ship, it may consist of a lifeboat ready for immediate use, preferably at the after end of the ship.

(30) Are sufficient personnel onboard and ashore to deal with an emergency?

At all times during the ship's stay at a terminal, a sufficient number of personnel should be present onboard the ship and in the shore installation to deal with an emergency.

(31) Are adequate insulating means in place in the ship/shore connection?

Unless measures are taken to break the continuous electrical path between ship and shore pipework provided by the ship/shore hoses or metallic arms, stray electric currents, mainly from corrosion prevention systems, can cause electric sparks at the flange faces when hoses are being connected and disconnected. The passage of these currents is usually prevented by an insulating flange inserted at each jetty manifold outlet or incorporated in the construction of metallic arms. Alternatively, the electrical discontinuity may be provided by the inclusion of one length of electrically discontinuous hose in each hose string. It should be ascertained that the means of electrical discontinuity is in place, is in good condition, and that it is not being bypassed by contact with an electrically conductive material.

(32) Have measures been taken to ensure sufficient pump room ventilation?

Pump rooms should be mechanically ventilated and the ventilation system, which should maintain a safe atmosphere throughout the pump room, should be kept running throughout the operation.

(33) If the ship is capable of closed loading, have the requirements for closed operations been agreed?

It is a requirement of many terminals that when the ship is ballasting, loading, and discharging, it operates without recourse to opening ullage and sighting ports. Such ships will require the means to enable closed monitoring of tank contents, either by a fixed gauging system or by using portable equipment passed through a vapour lock, and preferably backed up by an independent overfill alarm system.

(34) Has a vapour return line been connected? If required, a vapour return line may have to be used to return flammable vapours from the cargo tanks to shore.

(35) If a vapour return line is connected, have operating parameters been agreed upon?

The maximum and minimum operating pressures and any other constraints associated with the operation of the vapour return system should be discussed and agreed upon by ship and shore personnel.

(36) Are ship emergency fire control plans located externally?

A set of fire control plans should be permanently stored in a prominently marked weather-tight enclosure outside the deck house for the assistance of shore side firefighting personnel. A crew list should also be included in this enclosure.

(37) If the ship is fitted, or required to be fitted, with an inert gas system, the following questions should be answered:

(38) Is the inert gas system fully operational and in good working order?

The inert gas system should be in safe working condition with particular reference to all interlocking trips and associated alarms, deck seal, non-return valve, pressure regulating control system, main deck inert gas line pressure indicator, individual tank inert gas valves (when fitted), and deck P/V breaker. Individual tank inert gas valves (if fitted) should have easily identified and fully functioning open/close position indicators.

(39) Are deck seals in good working order?

It is essential that the deck seal arrangements are in a safe condition. In particular, the water supply arrangements to the seal and the proper functioning of associated alarms should be checked.

(40) Are liquid levels in P/V breakers correct?

Checks should be made to ensure the liquid level in the P/V breaker complies with manufacturer's recommendations.

(41) Have the fixed and portable oxygen analysers been calibrated and are they working properly?

All fixed and portable oxygen analysers should be calibrated and checked as required by the company and/or manufacturer's instructions. The in-line oxygen analyser/recorder and sufficient portable oxygen analysers should be working properly.

(42) Are fixed inert gas pressure and oxygen content recorders working?

All recording equipment should be switched on and operating correctly.

(43) Are all cargo tank atmospheres at positive pressure with an oxygen content of 8% or less by volume?

Prior to commencement of cargo operations, each cargo tank atmosphere should be checked to verify an oxygen content of 8% or less by volume. Inerted cargo tanks should at all times be kept at a positive pressure.

(44) Are all the individual tank inert gas valves (if fitted) correctly set and locked?

For both loading and discharge operations, it is normal and safe to keep all individual tank inert gas supply valves (if fitted) open in order to prevent inadvertent under- or over-pressurisation. In this mode of operation, each tank pressure will be the same as the deck main inert gas pressure and thus the P/V breaker will act as a safety valve in case of excessive over- or under-pressure. If individual tank inert gas supply valves are closed for reasons of potential vapour contamination or depressurisation for gauging, etc., then the status of the valve should be clearly indicated to all those involved in cargo operations. Each individual tank inert gas valve should be fitted with a locking device under the control of a responsible officer.

(45) Are all the persons in charge of cargo operations aware that in the case of failure of the inert gas plant, discharge operations should cease, and the terminal be advised?

In the case of failure of the inert gas plant, the cargo discharge, de-ballasting, and tank cleaning should cease and the terminal to be advised. Under no circumstances should the ship's officers allow the atmosphere in any tank to fall below atmospheric pressure. Section 10 of the IMO publication entitled 'Crude Oil Washing Systems' contains operational checklists for the use of the crew at each discharge in accordance with Regulation 13B of Annex I to MARPOL 73/78. If the ship is fitted with a crude oil washing (COW) system, and intends to crude oil wash, the following questions should be answered:

(45) (a) Is the Pre-Arrival Crude Oil Washing Checklist, as contained in the approved Crude Oil Washing Manual, satisfactorily completed?

The approved Crude Oil Washing Manual contains a Pre-Arrival Crude Oil Washing Checklist, specific to each ship, which should be completed by a responsible ship's officer prior to arrival at every discharge port where crude oil washing is intended.

(b) Is the Crude Oil Washing Checklist for use before, during, and after Crude Oil Washing, as contained in the approved Crude Oil Washing Manual, available and being used?

The approved Crude Oil Washing Manual contains a Crude Oil Washing Checklist, specific to each ship, for use before, during, and after crude oil washing operations. This Cheek List should be completed at the appropriate times and the terminal representative should be invited to participate.

8.4 SHIP/SHORE SAFETY CHECKLIST

Figure 8.2 is an example of a typical ship/shore safety checklist, which has been reproduced with kind permission of the Port of Helsingborg, Sweden.

In this chapter, we have examined some of the safety issues and hazards involved in the carriage of liquefied gases in bulk. The dangers associated with petroleum and natural gases cannot be overstated and should never be underestimated. Always follow the safety precautions and adhere to loading

14.0 SHIP-SHORE SAFETY CHECK LIST AND GUIDELINES*

Name of ship:

Berth: Port:

Date of arrival: Time of arrival:

INSTRUCTIONS FOR COMPLETION: The safety of operations requires that all questions should be answered affirmatively by clearly ticking (4) the appropriate box. If an affirmative answer is not possible, the reason should be given and agreement reached upon appropriate precautions to be taken between the ship and the terminal. Where any question is considered to be not applicable, then a note to that effect should be inserted in the remarks column.

A box in the columns 'ship' and 'terminal' indicates that checks should be carried out by the party concerned.

The presence of the letters **A, P** or **R** in the column 'Code' indicates the following:

A – any procedures and agreements should be in writing in the remarks column of this Check List or other mutually acceptable form. In either case, the signature of both parties should be required. **P** – in the case of a negative answer, the operation should not be carried out without the permission of the Port Authority. **R** – indicates items to be re-checked at intervals not exceeding that agreed in the declaration.

Part A - Bulk Liquid - General - Physical Checks	Ship	Terminal	Code	Remarks	
1. There is safe access between the ship and shore.	☐	☐	R		
2. The ship is securely moored.	☐	☐	R		
3. The agreed ship/shore communication system is operative.	☐	☐	AR	System:	Back up system:
4. Emergency towing-off pennants are correctly rigged and positioned.	☐	☐	R		
5. The ship's fire hoses and fire fighting equipment is positioned and ready for immediate use.	☐		R		
6. The terminal's fire-fighting equipment is positioned and ready for immediate use.		☐	R		
7. The ship's cargo and bunker hoses, pipelines and manifolds are in good condition, properly rigged and appropriate for the service intended.	☐				
8. The terminal's cargo and bunker hoses/arms are in good condition, properly rigged and appropriate for the service intended.		☐			
9. The cargo transfer system is sufficiently isolated and drained to allow safe removal of blank flanges prior to connection.	☐	☐			
10. Scuppers and 'save alls' on board are effectively plugged and drip trays are in position and empty.	☐		R		
11. Temporarily removed scupper plugs will be constantly monitored.	☐		R		
12. Shore spill containment and sumps are correctly managed.		☐	R		
13. The ship's unused cargo and bunker connections are properly secured with blank flanges fully bolted.	☐				
14. The terminal's unused cargo and bunker connections are properly secured with blank flanges fully bolted.		☐			
15. All cargo, ballast and bunker tank lids are closed.	☐				
16. Sea and overboard discharge valves, when not in use, are closed and visibly secured.	☐				
17. All external doors, ports and windows in the accommodation, stores and machinery spaces are closed. Engine room vents may be open.	☐		R		
18. The ship's emergency fire control plans are located externally.	☐			Location:	

*International Safety Guide for Oil Tankers and Terminal Fifht Edition 2006-12-05

Helsingborgs Hamn – Port of Helsingborg

Postal address	Harbour Office	Telephone	Email
Helsingborgs Hamn AB	Oceangatan 3	+46 (0)42 10 63 00	info@port.helsingborg.se
251 89 Helsingborg	Org No 5560240979	Telefax	Internet
SWEDEN	VAT No SE5560240979	+46 (0)42 12 43 74	www.port.helsingborg.se

Figure 8.2 Example of a ship/shore safety checklist.

Source: Port of Helsingborg, Email Authorisation.

☐ **HELSINGBORGS HAMN**
PORT OF HELSINGBORG

If the ship is fitted, or is required to be fitted with an Inert Gas System (IGS), the following points should be physically checked.

Bulk Liquid - General - Physical Checks	Ship	Terminal	Code	Remarks
19. Fixed IGS pressure and oxygen content recorders are working.	☐		R	
20. All cargo tank atmospheres are at positive pressure with oxygen content of 8% or less by volume.	☐		PR	

Part B - Bulk Liquid - General - Verbal Verification	Ship	Terminal	Code	Remarks
21. The ship is ready to move under its own power.	☐		PR	
22. There is an effective deck watch in attendance on board and adequate supervision of operations on the ship and in the terminal.	☐	☐	R	
23. There are sufficient personnel on board and ashore to deal with an emergency.	☐	☐	R	
24. The procedures for cargo, bunker and ballast handling have been agreed.	☐	☐	AR	
25. The emergency signal and shutdown procedure to be used by the ship and shore have been explained and understood.	☐	☐	A	
26. Material safety data sheets (MSDS) for the cargo transfer have been exchanged where requested.	☐	☐		
27. The hazards associated with toxic substances in the cargo being handled have been identified and understood.	☐	☐		H2S Content: Benzene Content:
28. An International Shore Fire Connection has been provided.	☐	☐		
29. The agreed tank venting system will be used.	☐	☐	AR	Method:
30. The requirements for closed operations have been agreed.	☐	☐	R	
31. The operation of the P/V system has been verified.	☐			
32. Where a vapour return line is connected, operating parameters have been agreed.	☐	☐	AR	
33. Independent high level alarms, if fitted, are operational and have been tested.	☐			
34. Adequate electrical insulating means are in place in the ship/shore connection.		☐		
35. Shore lines are fitted with a non return valve or procedures to avoid 'back filling' have been discussed.		☐		
36. Smoking rooms have been identified and smoking requirements are being observed.	☐	☐		Nominated smoking rooms:
37. Naked light regulations are being observed.	☐	☐	AR	
38. Ship/shore telephones, mobile phones and pager requirements are being observed.	☐	☐	AR	Location:
39. Hand torches (flashlights) are of an approved type.	☐	☐		
40. Fixed VHF/UHF transceivers and AIS equipment are on the correct power mode or switched off.	☐			
41. Portable VHF/UHF transceivers are of an approved type.	☐	☐		
42. The ship's main radio transmitter aerials are earthed and radars are switched off.	☐			
43. Electric cables to portable electrical equipment within the hazardous area are disconnected from power.	☐	☐		

Helsingborgs Hamn – Port of Helsingborg

Postal address
Helsingborgs Hamn AB
251 89 Helsingborg
SWEDEN

Harbour Office
Oceangatan 3
Org No 5560240979
VAT No SE5560240979

Telephone
+46 (0)42 10 63 00
Telefax
+46 (0)42 12 43 74

Email
info@port.helsingborg.se
Internet
www.port.helsingborg.se

Figure 8.2 (Continued)

HELSINGBORGS HAMN
PORT OF HELSINGBORG

	Ship	Terminal	Code	
44. Window type air conditioning units are disconnected.	☐			
45. Positive pressure is being maintained inside the accommodation.	☐			
46. Measures have been taken to ensure sufficient mechanical ventilation in the pump room.	☐		R	
47. There is provision for an emergency escape.	☐	☐		
48. The maximum wind and swell criteria for operations has been agreed.	☐	☐	A	Stop cargo at: Disconnect at: Unberth at:
49. Security protocols have been agreed between the Ship Security Officer and the Port Facility Security Officer, if appropriate.	☐	☐	A	

If the ship is fitted, or required to be fitted, with an Inert Gas System (IGS) the following statements should be addressed.

Inert Gas System	Ship	Terminal	Code	Remarks
50. The IGS is fully operational and in good working order	☐		R	
51. Deck seals, or equivalent, are in good working order.	☐		R	
52. Liquid levels in pressure/vacuum breakers are correct.	☐		R	
53. The fixed and portable oxygen analysers have been calibrated and are working properly.	☐		R	
54. All the individual tank IGS valves (if fitted) are correctly set and locked.	☐		R	
55. All personnel in charge of cargo operations are aware that in the case of failure of the Inert Gas Plant, discharge operations should cease, and the terminal be advised.	☐			

If the ship is fitted with a crude oil washing (COW) system, and intends to COW, the following statements should be addressed.

Crude Oil Washing	Ship	Terminal	Code	Remarks
56. The Pre-Arrival COW checklist, as contained in the approved COW manual, has been satisfactorily completed.	☐			
57. The COW check lists for use before, during and after COW, as contained in the approved COW manual, are available and being used.	☐		R	

Helsingborgs Hamn – Port of Helsingborg

Postal address	Harbour Office	Telephone	Email
Helsingborgs Hamn AB	Oceangatan 3	+46 (0)42 10 63 00	info@port.helsingborg.se
251 89 Helsingborg	Org No 5560240979	Telefax	Internet
SWEDEN	VAT No SE5560240979	+46 (0)42 12 43 74	www.port.helsingborg.se

Figure 8.2 (Continued)

HELSINGBORGS HAMN
PORT OF HELSINGBORG

the ship is fitted with a crude oil washing (COW) system, and intends to COW, the following statements should be addressed.

Tank Cleaning	Ship	Terminal	Code	Remarks
58. The Tank cleaning operations are planned during the ship's stay alongside the shore installation.	Yes / No* ☐ ☐	Yes / No* ☐ ☐		
59. If 'yes' the procedures and approvals for tank cleaning have been agreed.	☐	☐		
60. Permission has been granted for gas freeing operations.	Yes / No* ☐ ☐	Yes / No* ☐ ☐		

* Yes or No as appropriate

Part C - Bulk Liquid Chemicals - Verbal Verifications	Ship	Terminal	Code	Remarks
1. Material Safety Data Sheets are available giving the necessary data for the safe handling of the cargo.	☐	☐		
2. A manufacturer's inhibition certificate, where applicable, has been provided.	☐	☐	P	
3. Counter measures against accidental personal contact with the cargo have been agreed	☐	☐		
4. Sufficient protective clothing and equipment (including self-contained breathing apparatus) is ready for immediate use and is suitable for the product being handled.	☐	☐		
5. The cargo handling rate is compatible with the automatic shut down system, if in use.	☐	☐	A	
6. Cargo system gauges and alarms are correctly set and in good order.	☐	☐		
7. Portable vapour detection instruments are readily available for the products being handled.	☐	☐		
8. Information on fire-fighting media and procedures has been exchanged.	☐	☐		
9. Transfer hoses are of suitable material, resistant to the action of the products being handled.		☐		
10. Cargo handling is being performed with the permanent installed pipeline system.	☐	☐		

Part D - Bulk Liquid Chemicals - Verbal Verifications	Ship	Terminal	Code	Remarks
1. Material Safety Data Sheets are available giving the necessary data for the safe handling of the cargo..	☐	☐		
2. A manufacturer's inhibition certificate, where applicable, has been provided	☐	☐	P	
3. The water spray system is ready for immediate use.	☐	☐		
4. There is sufficient protective equipment (including self-contained breathing apparatus) and protective clothing ready for immediate use..	☐	☐		
5. Hold and inter-barrier spaces are properly inerted or filled with dry air, as required.	☐			
6. All remote control valves are in working order.	☐	☐		
7. The required cargo pumps and compressors are in good order, and the maximum working pressures have been agreed between ship and shore.	☐	☐	A	
8. Re-liquefaction or boil off control equipment is in good order.	☐	☐		

Helsingborgs Hamn – Port of Helsingborg

Postal address
Helsingborgs Hamn AB
251 89 Helsingborg
SWEDEN

Harbour Office
Oceangatan 3
Org No 5560240979
VAT No SE5560240979

Telephone
+46 (0)42 10 63 0
Telefax
+46 (0)42 12 43 74

Email
info@port.helsingborg.se
Internet
www.port.helsingborg.se

Figure 8.2 (Continued)

9. The gas detection equipment has been properly set for the cargo, is calibrated and is in good order.	☐	☐			
10. Cargo system gauges and alarms are correctly set and in good order.	☐	☐			
11. Emergency shutdown systems have been tested and are working properly.	☐	☐			
12. Ship and shore have informed each other of the closing rate of ESD valves, automatic valves or similar devices.	☐	☐	A	Ship: Shore:	
13. Information has been exchanged between ship and shore on the maximum/minimum temperatures/pressures of the cargo to be handled.	☐	☐	A		
14. Cargo tanks are protected against inadvertent overfilling at all times while any cargo operations are in progress.	☐	☐			
15. The compressor room is properly ventilated; the electrical motor room is properly pressurised, and the alarm system is working.	☐				

DECLARATION

We, the undersigned, have checked the above items in Parts A and B, and where appropriate, Part C or D, in accordance with the instructions and have satisfied ourselves that the entries we have made are correct to the best of our knowledge. We have also made arrangements to carry out repetitive checks as necessary and agreed that those items coded 'R' in the Check List should be re-checked at intervals not exceeding   hours.
If to our knowledge the status of any item changes, we will immediately inform the other party.

For Ship	For Shore
Name:	Name:
Rank:	Rank:
Signature:	Signature:
Date:	Date:
Time:	Time:

Record of repetitive checks:

Date:								
Time:								
Initials for Ship:								
Initials for Shore:								

Helsingborgs Hamn – Port of Helsingborg

Postal address	Harbour Office	Telephone	Email
Helsingborgs Hamn AB	Oceangatan 3	+46 (0)42 10 63 00	info@port.helsingborg.se
251 89 Helsingborg	Org No 5560240979	Telefax	Internet
SWEDEN	VAT No SE5560240979	+46 (0)42 12 43 74	www.port.helsingborg.se

Figure 8.2 (Continued)

and discharging procedures without fail or exception. In the next chapter, we will look at how various cargoes are loaded, carried, and discharged from offshore support and supply vessels. Due to the unusual and unique design, these types of vessels have very specific methods for loading and discharging their cargoes. Furthermore, as they operate in some of the harshest maritime environments around the world, the manner with which cargoes are stowed is also discussed in detail.

Chapter 9

Principles for the Safe Handling, Stowage, and Carriage of Cargo on Offshore Supply Vessels

9.1 INTRODUCTION

Offshore supply vessels are often said to be the most dangerous when it comes to cargo operations (Figure 9.1). The most common types of cargo carried by these special-purpose vessels are containers (of different sizes and heights), chemicals, explosives, radioactive materials, large tubulars, and heavy lifts of irregular shapes and sizes. The cargo is loaded on deck and, since the vessel has a low freeboard, and usually faces severe weather, handling and securing of the cargo is of paramount importance. Ideally, the deck crew should be able to work the cargo on the deck without having to move between the containers and other stowed cargoes. This means careful attention to the security and stowage of cargo is essential on offshore supply vessels. Usually, block stowage of container cargo is a good form of restraining cargo athwartships movement. To ensure safety, the following information is required at the preplanning stage:

(1) The names and limitations of the installation(s) to be called at
(2) The order in which the installation(s) are to be called at
(3) Any items for urgent delivery
(4) Any heavy lifts and their individual weights
(5) Identification of dangerous goods
(6) Numbers and sizes of tubular steel pipes
(7) The order and extent of any back loading to be handled at each installation

Another type of cargo that is typically carried onboard offshore supply vessels are dangerous goods and pollutants. The vessel's master must take every possible precaution to prevent any pollution or incident. For this, the importance of securing and segregation is again emphasised. Tubular steel pipes are prone to athwartships movement. Stowing them on a bed of old ropes on deck, then chaining them down into place, can help restrict this. Tugger winches are used to tighten stows prior to chaining. These vessels also have

DOI: 10.1201/9781003354338-10

Figure 9.1 Offshore supply vessel – *MV Havila Clipper.*

Source: Author's own.

a variety of tanks which are used to carry bulk cargoes such as cement and lubricating oil. These commodities are handled through hoses, which means that segregation is very important. Bulk tanks should be cleaned and dried prior to the commencement of loading. Extreme care should be exercised when gas oil is loaded, transferred, and carried onboard offshore supply vessels. If heavy weather is expected, consideration must be given to any possible loss of cargo through the tank vents, thus causing pollution. A current cargo plan must be maintained.

9.2 PREPARING TO LOAD CARGO ONBOARD

When planning to load any cargo, on or under deck, it is the joint responsibility of charterer, owner, master, and base operator to ensure that the vessel is fit for purpose and in compliance with all relevant requirements relating to the safe carriage of the goods or products. This includes compliance with relevant international regulations as well as the rules and codes of the vessel's Flag State and those of the regional authorities in the present area of operation. The charterer, owner, base operator, and master should ensure that all personnel who are involved in the loading or discharge of cargo are qualified and competent in the handling and carriage of the goods or products on offshore vessels. This requirement also extends to any other personnel who may be mobilised to provide support services which might be necessary, including surveyors and other quality assurance specialists. Whilst these responsibilities relate particularly to the carriage of dangerous goods and inflammable, noxious, or otherwise hazardous liquid products, they also relate to all other cargoes carried on offshore supply vessels.

9.3 NOTIFICATION OF ANY UNUSUAL CARGO ITEMS

Where there is an intention to ship any unusual cargo items on an offshore supply vessel, the base operator must advise the vessel's master in a timely manner in order that any risks associated with the shipment can be properly assessed and appropriate preparations made. Items falling into this category are referred to in sections 7.3.2 (Certain Lifting Operations) and 9.13 (Unusual Cargo Items Loaded onto Vessel Decks) of the Guidelines for Offshore Marine Operations (GOMO).

9.4 DECK SPACE MANAGEMENT AND BACK LOAD CARGOES

Congestion on the cargo decks of both vessel and offshore facilities can result in the development of situations hazardous to personnel or equipment. Except where rigorous planning of logistics support is in place or where previously agreed upon and confirmed in sailing instructions, it is considered good practice for the vessel to arrive at the offshore facility with approximately 10% of its usable deck clear and ready to receive any initial backload. This allows sufficient space to be cleared on the facility's deck before any cargo is brought up from the vessel. Wherever possible, this clear deck space should be contiguous. Subject to discussion with the master, this recommendation may be waived at the last facility at which cargo is backloaded onto the vessel prior to its return to base when all deck space may be utilised, but only on the understanding that it will not subsequently be diverted to support another offshore location on its inward voyage (Figure 9.2).

9.5 CARGO PLANS

In the course of the initial loading at its shore base(s), the master should ensure that a record of the cargo loaded onboard is maintained. This should show the locations of the 'blocks' of cargo for each facility to be supported

Figure 9.2 Offshore containers.

Source: Author's own.

during the forthcoming voyage, together with the number of lifts in each block and any other relevant details. Locations of any unusual cargo items should be clearly identified. The cargo plan may be further supported by photographs of the vessel's deck. There is normally a requirement for this plan to be forwarded to the base operator on completion of loading, who will subsequently arrange for it to be forwarded to the offshore facility(ies). The plan should be updated as the voyage progresses. A typical deck plan is illustrated herein. Other examples, based on electronic software packages exist, and may be more easily transmitted through the usual onboard communications channels. A table or drawing showing the contents of the vessel's underdeck cargo tanks should also be prepared and forwarded to the base operator as described earlier.

9.6 SAILING INSTRUCTIONS

Prior to the vessel being dispatched on any voyage to deliver cargoes to one or more offshore facilities, the base operator, in conjunction with the charterer, should furnish the vessel with a comprehensive set of sailing instructions. These instructions may include, but are not limited to, any of the following:

(1) The cargo manifest (which includes details of items loaded on the vessel).
(2) Any specific information regarding cargoes onboard, including (MSDS) and specific hazards associated with any cargoes.
(3) Any specific precautions relating to the care of any cargo.
(4) Passage planning (routing) for the voyage.
(5) Facility data cards, if not already held onboard.
(6) Reporting requirements.
(7) Any changes in contact details.
(8) Any other special instructions or relevant information.

Arrangements should be made with a reputable weather forecasting service provider, experienced in the preparation of offshore forecasts, to prepare and promulgate weather forecasts extending, where practicable, up to five days (i.e. 120 hours) for the relevant locations. These forecasts are generally arranged by the charterer and should be made available to the vessel master. The weather forecasting service provider may also be able to prepare more specialised information on request, including longer term forecasts, met-ocean statistical analyses, and so forth. It is also the master's responsibility to ensure that forecasts from other publicly available sources can be received onboard and taken into account during passage planning. Where the forecasts received indicate prolonged periods of adverse weather at the offshore locations to be supported on a particular voyage such that it is unlikely that any of the intended sites can be worked safely, the master and charterer should agree that the dispatch of any vessels involved should be deferred until the weather conditions improve. In the event that a vessel is dispatched in such circumstances, the master may, at their sole discretion, elect to take an indirect route to reduce the risks to the ship, its personnel, and cargo, or to proceed to a sheltered location to await an improvement in conditions at the offshore location. In this context, 'prolonged period' should be taken as any period exceeding one day (any 24-hour period or part thereof) where it is unlikely that any work can be safely undertaken at the offshore location.

9.7 POTENTIAL DROPPED OBJECTS

Unsecured objects being dislodged or falling from cargo items represent a serious risk to personnel, equipment, and the environment. This is a risk that exists throughout the logistics supply chain. At all stages in the supply chain, items should therefore be thoroughly inspected prior to transfer from one stage to the next. Potential dropped objects identified during these inspections should be removed and reported accordingly. Objects which constitute a dropped objects risk include, but are not limited to, loose tools used when servicing equipment included in or forming part of the cargo item; foreign objects in or on containers, including forklift pockets; and ice formed when water entrained in a cargo item freezes. When loading or discharging, any deck cargo personnel involved in the procedure should position themselves within a safe haven that is well and clear of the intended load path until such time as it is safe to approach the item. Where practical and safe to do so, items on deck should be inspected for potential dropped objects after loading and again before discharge at the offshore facility or onshore base. Any such objects identified during these inspections should be removed if safe to do so and an incident report submitted through the appropriate means. If the object(s) cannot be safely removed, the cargo item should be quarantined pending a full assessment of the risks which may be involved in discharge either at the offshore facility or at the quayside (Figure 9.3).

9.8 STOWAGE AND SECURING OF CARGOES IN CONTAINERS

Failure to correctly secure cargo items shipped in containers, whether open or closed, can present serious risks to personnel and equipment, including injuries sustained by crew members when attempting to secure loose items. This is further increased by changes in the centre of gravity of the lift due to the movement of loose items within the container. This can result in the contents of the container falling significantly out of level. This may result in the loss of contents from the container. Handling of loads, particularly when landing, will also be made much more difficult. The proper packing and securing of cargo within containers are therefore a safety matter of the utmost significance. Any person who has reason to believe that the correct

Figure 9.3 Offloading offshore containers.

Source: Author's own.

Figure 9.4 Offshore containers.

Source: Author's own.

procedures have not been followed, or where satisfactory arrangements are not in place, should 'stop the job' immediately until remedial measures have been implemented (Figure 9.4).

9.9 REFRIGERATED CONTAINERS (DISCONNECTION AT THE OFFSHORE FACILITY)

From time-to-time refrigerated containers may be used to deliver provisions to offshore facilities. Such containers may have their own self-contained refrigeration unit, but more usually electrically powered units will require connection to receptacles on vessels which have been specifically installed for this purpose. Specific checklists may relate to the carriage of such items which should be completed by the relevant personnel. Where such container(s) are used, it is important that they are not isolated for significant periods since the temperature may rise to such an extent that the contents thaw and must be condemned. It is therefore recommended that when preparing to discharge this type of container at the offshore facility the power supply is isolated, disconnected, and removed only from those items to be delivered at that facility. The power supply may remain connected to refrigerated containers intended for other destinations. In some circumstances, it may be necessary to isolate, disconnect, and remove the power supply to those containers to be delivered to an offshore facility prior to entering its safety zone (Figure 9.5).

Figure 9.5 Preparing to load offshore containers.

Source: Author's own.

9.10 TUBULAR CARGO

For round-tripped tubular cargo, it is recommended that when tubular cargoes remain on the vessel for successive voyages to an offshore facility, the following practices are adopted to prevent incidents from occurring (Figure 9.6). Lifting arrangements should be checked to ensure that they are correctly installed prior to loading any other similar items over the top. Such checks should include the following:

(1) Correct leads of all parts of lifting arrangements.
(2) Correct installation of securing arrangements such as bulldog grips, Velcro straps, tie-wraps, etc.
(3) An assessment of the adequacy and suitability of the aforementioned securing arrangements.

Prior to lifting bundles from the vessel deck to the offshore facility, a check should be made of *both* ends of the lifting slings to ensure that they are correctly set up for the lift. Where appropriate, a risk assessment of the discharge of such items should be undertaken and the outcomes included for discussion in the toolbox talk.

Figure 9.6 Example of typical tubulars.

Source: Author's own.

9.11 MAIN BLOCK OPERATIONS

Cargo items will normally be transferred between the vessel's deck and the offshore facility using the auxiliary hoist (the *whip line*) of the facilities derrick. From time to time, however, where the weight of the item to be transferred exceeds the capacity of the auxiliary hoist, the derrick's main hoist must be used. Should this be necessary, an intermediate pennant of sufficient safe working load should be installed on the hook(s) of the main block enabling personnel to connect or release the lifting rigging on the cargo item without having to approach or attempt to manoeuvre the block itself. Where practical, this intermediate pennant should be of sufficient length such that the height of the main block when the lifting rigging is connected or released is always approximately 5 metres (16 ft) above the cargo rails at the side of the main deck, or the highest adjacent item of cargo if this extends above the cargo rail. Any requirements to undertake operations of this nature should be advised to the vessel(s) involved in sufficient time for the appropriate task-specific risk assessments to have been made. Operations should not commence until the vessel has confirmed that these assessments have been completed and personnel briefed as to any precautions to be observed.

9.12 CHERRY PICKING

'Cherry Picking' is defined the 'selective discharge of cargo from within the stow'. The term 'cherry picking' generally infers cargo lifting arrangements not being directly accessible from the deck level. This usually involves breaking the stow from an open location with no clear and secure access or escape route(s) to adjacent safe havens. It may also require personnel to use unsecured ladders, or to climb on top of other cargo or even the ship's structure to enter containers. This process for connecting lifting arrangements is strictly prohibited at all times. Masters who may be asked to undertake any of the above activities should 'stop the job' immediately. To minimise the risk of 'cherry picking', every effort should be made prior to the start of loading to ascertain which, if any, cargo items are of high priority. Vessels should be advised accordingly with cargo stowed in such a manner that any high-priority items can be discharged directly on arrival at the destination. These cargoes should be identified *before* the cargo is loaded onto the vessel.

Other potentially dangerous practices include moving other cargo on deck of vessel to gain access to a particular item. Lifting cargo containers to the deck of the facility, stripping the same and returning to the vessel with the vessel being required to remain on station at the installation throughout. It should be appreciated that such practices may introduce increased risks due to additional lifting operations, involving increased risk to personnel; or vessels having to remain in close proximity or even adjacent to the offshore facility. Doing so present a heightened risk of collision. Masters who are requested to undertake either of the above should challenge such requests, drawing attention to the additional risks outlined earlier. Furthermore, before proceeding with any of the activities referred to earlier, a thorough risk assessment should be undertaken and the outcomes included for discussion in any subsequent toolbox talks. Where frequent requests to support operations of this nature are received from a facility, concerns relating to the risks involved should be raised with the facility manager and the charterer.

9.13 UNUSUAL CARGO ITEMS LOADED ONTO VESSEL DECKS

From time-to-time requirements may exist for unusual items to be loaded onto the vessel's deck. Examples of such items include but are not limited to modules or large fabricated items associated with offshore construction projects, unusually long items such as tubulars, flare booms, and crane booms, which because of lifting geometry require the use of two stinger pennants on the crane hook, and any items which have not been pre-slung prior to shipment. Such items may have unusual dimensions, be unduly heavy or have high footprint loads, have unusual means of support, which means their transportation may have been the subject of a specific engineering assessment. In addition, the connection and release of the lifting rigging may pose specific risks for the ship's staff. In this context, any cargo items not carried in conventional shipping units such as containers, baskets, tanks, or racks should be considered 'unusual'. The vessel master should be notified in advance of the intention to load such items onto their vessel with sufficient time to perform a detailed risk assessment. In general, it is recommended that the use of tag lines should be avoided. However, it is recognised that tag lines may be advantageous in handling some of the cargo items referred to earlier, also that they are in general use in certain parts of the world.

9.14 BULK CARGO GENERAL REQUIREMENTS

Cargoes carried in bulk on offshore supply vessels include dry products in powder form together with various types of oil- and water-based muds, base oils, brine, and numerous other chemicals that are transported in liquid form. When undertaking bulk cargo operations, the following precautions should be observed:

(1) The pressure ratings of all components of the transfer system should be verified to ensure that they are appropriate for the proposed operation.
(2) Prior to commencement, agreement should be reached between all relevant parties including the vessel, base, offshore facility, or roadside tanker regarding the pressure rating to avoid overpressure.
(3) The protocols for control of the transfer operation are to be agreed upon in advance by all concerned.
(4) Communications arrangements should be agreed upon and tested prior to the start of operations and at frequent intervals thereafter.
(5) If communications are lost, *stop the job*.
(6) The shipper and receiver should confirm the quantities to be transferred and thereafter monitor at regular intervals.
(7) The shipper and receiver should agree the rates of delivery and the densities of cargo to be transferred.
(8) All relevant ship's staff and facility personnel should be readily available and positioned nearby throughout transfer operation.
(9) At the facility, the master or senior OOW must ensure they can always see the bulk hose(s). It is imperative they are not distracted in any way.

(10) Particular attention should be paid during hydrocarbon transfers that proper consideration is given to potential hazards when carrying out concurrent cargo operations. Each party must provide sufficient warning prior to changing over tanks and communicate when changes have occurred.

If at any point the vessel master, shipper, offshore installation manager (OIM), or any other person has concerns relating to the safety of the transfer, the operation must be terminated immediately. Unregulated compressed air should never be used to clear the bulk hoses which are fed back to the vessel as this may damage the tanks. Compressed air should not be used to clear hoses used for the transfer of any hydrocarbon-based products as there is an increased risk of explosion. It is important to note that any liquids not of a hydrocarbon nature should be transferred using potable water hoses. Before use, the potable water hose(s) should be flushed thoroughly with potable water lines to clear any residues. Hoses should always remain afloat through sufficient floating devices. The use of self-sealing weak link couplings in the mid-section of the hose string is strongly recommended. Always avoid the use of heavy sections of reducers or connections at the hose ends. The hose from the facility should not be connected to the vessel until both the vessel and the facility have agreed that all preparations are complete and that the transfer can commence immediately after the final connection has been satisfactorily made.

9.15 BULK OPERATIONS IN PORT AND AT THE FACILITY

Specific responsibilities associated with bulk operations in port and at the facility are described in the following sections. A checklist which should be completed prior to commencing any transfers of bulk cargoes is also included herein.

9.15.1 Vessel Responsibilities at the Facility

Before offloading bulk cargo, it is necessary for the vessel and the offshore facility to confirm a number of key parameters, including communication protocols with the receiving facility; whose 'STOP' it is; the quantity of bulk to be offloaded; the necessary hoses and connections, colour codes, and dimensions; whether the rigged hose lengths are adequate; the procedures for venting and blowing through the hoses; whether the facility is ready to receive the cargo; whether all valves and vents are open and the correct tanks are lined up; and whether the relevant emergency shutdown procedures are in place and the crew are familiar with same. It is also necessary to ensure that all pollution prevention equipment is in place as per the

SMPEP and all manifold valves must be in good condition. Moreover, the PIC cannot be distracted from the operation and any facility under-deck lighting must be adequate. Dry bulk vent line positions must be identified and, where necessary, prepared. The master should submit to the designated contact person all receipts where applicable, including meter-slips, for the cargoes transferred in addition to any other relevant documentation and information.

9.15.2 Facility Responsibilities

The facility must ensure that communication protocols have been agreed upon and in particular whose 'STOP' it is. Hoses, manifolds and valves must be visually inspected, maintained, and replaced as required and/or in accordance with the planned maintenance system. Slings and lifting points must be visually checked and replaced as required. Hoses should only be lifted by a certified wire strop on a certified hook eye fitting. Under-deck lighting must adequately illuminate the transfer hose and vessel. Appropriate flotation systems must be intact and in place.

9.16 PREPARATIONS RELATING TO THE TRANSFER OF DRY BULK MATERIALS

It is recommended that procedures should be adopted as follows. Prior to confirming that the vessel is ready to transfer dry bulk cargoes, it should be verified that all onboard preparations have been completed. This includes a requirement to ensure that, where relevant, all elements of the system have been vented to atmospheric pressure. When transferring dry bulk cargoes to or from vessels, the personnel responsible for delivering the product should confirm that those responsible for receiving it have completed all relevant preparations. Assumptions that preparations have been completed can be dangerous and must be avoided at all costs. Relevant checklists are to be completed as required by the parties involved. When transferring dry bulk cargoes to or from vessels, care should be taken when deciding the sequence and manner which the various valves are opened to avoid the risk of inadvertently over-pressurising any elements of the system. It will be appreciated that the handling of dry bulk materials involves systems containing large volumes of pressurised air. The stored energy in such systems is therefore considerable and the potential for serious personal injury in the event of failure is high. All personnel involved in such operations must therefore comply with all relevant procedures and to ensure that all checks have been satisfactorily completed prior to confirming their readiness to deliver or receive the product.

9.17 BULK HOSE-HANDLING PROCEDURES AT THE FACILITY

It is recommended that the following procedures be adopted during receipt and handling of bulk hoses at the offshore facility. The vessel should take up position and confirm readiness to receive the hose. Except where other arrangements are in use, the crane operator on the facility lowers the hose to the vessel, holding the hose against the ship's side and at a height that allows the crew to catch and secure it to the vessel's side rail, keeping the hose end clear of the crews' heads. Where other arrangements exist, the appropriate procedures should be followed. Once secure, the hose end is lowered inboard of the rail and the crane hook disconnected. When the hook is clear, the crew install the hose on the appropriate connection on the ship's manifold. Uncoupling is the reverse of the aforementioned procedure. After releasing any self-sealing connection, it should be visually inspected by the deck crew to ensure that it is fully closed and is not passing any liquids. Vessel crews should be reminded that hose couplings should, whenever possible, avoid contact with the ship's structure. The integrity of the couplings should be monitored by visual inspection of the painted line on the couplings, where applied. In marginal weather, greater care than normal is needed by the vessel to avoid overrunning the hose, especially if deck cargo is also being worked. Consideration should therefore be given to working bulk only in such circumstances.

9.18 HOSE SECURING ARRANGEMENTS

Section 10.7 of the *Guidelines for Offshore Marine Operations* describes the general principles of handling bulk hoses at offshore facilities; however, these often require personnel securing the hose(s) to work underneath or in very close proximity to the suspended hose for some time whilst completing the arrangements used to secure the hose. Any arrangements which reduce the time personnel are required to work in proximity to the suspended hose or avoid this altogether should be investigated. When it is being passed to the vessel, personnel still work in proximity to the suspended hose when securing it, the time required is much reduced. On recovery, personnel are required to disconnect from the manifold and connect the crane hook to the recovery pennant, but thereafter further intervention is not normally required. Other arrangements having similar objectives have also been developed. These also involve minimal modifications to the vessel and have been used successfully in some operational areas. Whilst undoubtedly minimising risks associated with hose handling even further, such arrangements are likely to be more complex, requiring extensive modifications on both vessel and facility. The flexibility to utilise vessels not equipped with the features required will therefore be reduced.

9.19 BULK TRANSFERS OF COMMON LIQUIDS

There are a number of procedures that must be followed when preparing for, or carrying out bulk transfers of common liquids, which are briefly discussed here.

- *Cargo fuel (marine gas oil):* Establish a sampling and receipting procedure when transferring fuel. Sampling taken in accordance with *MARPOL Annex VI* will normally suffice for these operations. However, in some circumstances more rigorous sampling procedures may be required. Any such requirements should be included in the master's sailing instructions and should always be complied with.
- *Potable water:* Specific national or charterer's requirements may apply to the carriage, storage, and transfer of potable water. The charterer, owner, and master should ensure that any such requirements are understood and fully complied with.

9.20 BULK TRANSFERS OF SPECIAL PRODUCTS

Special care must be taken to follow the correct procedures when transferring special products, which include, but are not limited to, methanol and zinc bromide. Appropriate risk management procedures should be in place when transferring special products. When transferring these products, the following should be observed.

- *Shipper:* Provide full details of the product(s) being shipped, including details of all precautions to be taken when handling said products. Staff to be on site throughout to advise on pumping, handling, earthing, and the discharging of tanks. Provision of appropriate firefighting equipment, where relevant.
- *Operating company and base operator:* Nomination of berth after liaising with the harbour authority, fire brigade, and harbour police or security. Ensure sufficient cooling or drenching water is available. Cordon-off area, with signs posted to indicate a hazardous area.
- *Master:* Should complete a ship to shore safety check with the shipper. Must authorise loading. If required, ensure a permit to work is in place before any loading operations are performed. Ensure the vessel's restricted zone is clear, fire hoses are rigged, and SMPEP equipment is ready for action before commencing loading.

9.20.1 Characteristics of Some Special Liquid Products

Whilst the shipper should provide full details of any products being shipped, some of the main characteristics of the more common chemicals which may be shipped in bulk liquid form are included here.

- *Methanol:* Particular characteristics of this product are it burns with no visible flame in daylight conditions; is readily or completely miscible with water; is a class 3 substance with a noticeable odour; is highly flammable, with a flashpoint below 23°C (73.4°F); can evaporate quickly; has heavier than air vapour that may be invisible, and disperses over the ground; can form an explosive mixture with air, particularly in empty unclean offshore containers; experiences pressure increase on heating, with the risk of bursting followed by explosion; is very toxic, and possibly fatal, if swallowed or absorbed through the skin. Symptoms may not appear for several hours; and can cause significant irritation of the eyes. The following specific precautions should be observed when transferring methanol. Ensure that the integrity of the system is intact, including all relevant certification which should be valid and in-date. During bulk methanol transfer, smoking and the use of ignition sources are strictly prohibited. During electrical storms (lightning), operations should be terminated. Free deck space around bulk loading/discharge stations so that coverage of foam monitors is not obstructed. No other operations to be undertaken when handling this product.
- *Zinc Bromide:* Zinc bromide is a highly corrosive and environment-contaminating product. Due to its corrosive nature, protection against injury from exposure to it is essential. Information provided by the shipper should be used when undertaking risk assessments involving the carriage of this product to determine the appropriate level of personal protective equipment which should be used.

9.21 ATTENDANCE OF FACILITY PERSONNEL DURING BULK TRANSFER OPERATIONS

Whilst vessels are connected to offshore facilities by hose(s) for the purpose of delivering bulk commodities to facilities, it is important that, in the event of a change in the operating circumstances developing, personnel on the facility remain available at all times to disconnect the hose(s) at short notice. Failure to disconnect the hose(s) in a timely manner should circumstances change during bulk transfer operations could well result in significant risk of injury to personnel and/or damage to assets or the environment. The crane operator and deck crew on the facility shall therefore remain readily available, contactable, and nearby throughout transfer operations. In the event that any such personnel are required to leave the vicinity of operations for any reason, the vessel should be immediately advised. The vessel bridge team in conjunction with the facility manager should assess current and anticipated operational risks. It is the master's decision as to whether the vessel remains connected to the facility pending restoration of the required level of support.

9.22 TRANSFER OF NOXIOUS LIQUIDS DURING THE HOURS OF DARKNESS

It is recognised that it will be necessary to transfer hydrocarbon or other noxious liquids during the hours of darkness, particularly in higher latitudes in the winter months. For clarity, the Guidelines do not advocate that such operations should be curtailed or restricted, but instead seek to identify the additional risks involved in such transfers and to make appropriate recommendations to manage such risks. It is recognised, for example, that leaks are most likely to occur in the early phases of any transfer operation as connections become pressurised. Once all aspects of the transfer operation have been stabilised leaks are less likely to occur. It is therefore recommended that, wherever practical, the following practices may be adopted in relation to the bulk transfer of hydrocarbons (or other recognised marine pollutants) during the hours of darkness: adequate artificial illumination of the operational areas on the facility, the vessel and the water between them should be provided; additional high-visibility and/or reflective panels on the hoses (or their buoyancy elements) are recommended; all preparations for the transfer to be completed in daylight, where practical; careful check to be made for leaks, etc. on vessel, facility, and connecting hose as transfer commences; transfer may continue into the hours of darkness, provided that the entire area and associated equipment is adequately illuminated to an acceptable standard; in the event that the transfer continues a careful watch of the connections and hose should be maintained throughout; it is recommended that hydrocarbons or other noxious products should not be transferred simultaneously in these circumstances; on completion of the transfer, extra care should be taken when breaking the connection and returning the hose to ensure that the risk of spillage on completion of the operation is also minimised. General precautions to be observed regarding safety of personnel working on deck during the hours of darkness should continue to be implemented.

9.22.1 Tank Cleaning

The tank cleaning foreman must demonstrate to the master that they understand the principles and, if necessary, has undertaken a risk assessment relevant to the intended task. The outcomes of the risk assessment should have been addressed in the subsequent toolbox talk prior to commencing the task.

- *Personal protective equipment:* Personnel working in the tank must wear the appropriate personal protective equipment as identified in the risk assessment, COSHH, or equivalent assessment and MSDS.
- *Atmosphere Testing and tank entry:* All tanks should be considered as 'dangerous spaces' which, if appropriate precautions are not taken, would represent a serious risk to personnel entering them. The tank

cleaning foreman must demonstrate to the master that the atmosphere in the tank has been tested to prove that it does not represent a threat to any personnel who may be required to enter the space. They must also be able to demonstrate that any equipment utilised for this purpose has been used in accordance with the original equipment manufacturer's instructions. The results of the atmosphere testing should be recorded on the permit or other agreed document.

- *Communications.* Communication systems between all personnel within the tank and at the point of access must be agreed upon, tested prior to commencement of cleaning activities, and checked at frequent intervals until all persons have exited the tank on completion of operations. A standby person at each tank entrance will almost always be required. This person should be competent and trained to take the necessary action in the event of an emergency. Effective means of ship-to-ship and ship-to-shore communication must be established and maintained throughout the tank cleaning operation.
- *Emergency response and escape:* The tank cleaning foreman must demonstrate to the master that the emergency response and escape arrangements identified in the risk assessment are in place and available if required.
- *Checklist.* A typical example of a checklist which should be completed prior to the commencement of tank cleaning operations is included in Figures 9.7 and 9.8.

Figure 9.7 Tank cleaning.

Source: Author's own.

TANK MAINTENANCE CHECKLIST

Site: _____   Date: _____

Inspected By: _____

TANK CHECKS	OK	ACTION REQUIRED
Can you see any corrosion, rust, cracks, or warping on the tank body or bund?		
Can you see any corrosion, rust or leaks at the welded joints or the pipework?		
Is the tank base and any supporting structures in a good, safe condition?		
Is there evidence on the surrounding ground of fuel leaking e.g. damp spots?		
Is there any problematic water content or nasty looking sludge present in the tank?**		
Are all vents and fill points protected to prevent water and debris from entering?		
Does the bund need emptying of rainwater, leaves, fuel or any other waste?		
If there is a drip tray in the cabinet, does this need emptying of excess fuel?		
Is the tank labelling and any operating instructions still present and legible?		
Is the tank secured from unauthorised access and the locks maintained?		

**If water content is suspected, you should arrange for a fuel tank cleaning specialist to come out

EQUIPMENT CHECKS	OK	ACTION REQUIRED
Does the pump turn on and operate free of any unusual noises or problems?		
Are the hoses in good condition, free of leaks, splits and have not gone rigid?		
Is the meter calibrated correctly to count the fuel dispensed accurately?		
Have all filters been checked and replaced if almost full to prevent blocking?		
Is the level gauge reporting accurately and the reading clearly visible at the tank?		
Have any overfill, low level and bund alarms been tested to check they sound and operate in the correct manner?		
Do all isolation valves open and close as they should?		
Does the equipment look in good condition, clean and free from any leaks?		

Figure 9.8 Example of a typical tank cleaning checklist.

Source: Author's own.

9.22.2 Control

Although the tank cleaning operation is conducted by a contractor under control of the contractor's supervisor, the safety of the operation remains the responsibility of the master. The operation should be continuously monitored by a designated responsible vessel person who should stop any operation that they consider unsafe.

- *Atmosphere testing:* Regular tank atmosphere testing by competent personnel from both the vessel and the tank cleaning contractor must be undertaken prior to the commencement of cleaning activities and checked at frequent intervals until all persons have exited the tank on completion of operations. Equipment utilised to conduct these tests of the tank atmosphere must be used in accordance with the original equipment manufacturer's instructions.
- *Simultaneous operations.* Where simultaneous tank cleaning and other operations, that is, cargo operations, are undertaken, then suitable safety precautions must be in place. Interfaces between the vessel's officers, tank cleaning supervisors, and quay supervisors must be kept open and active throughout the tank cleaning operation.
- *Shift handovers.* Handover between shifts of the vessel's and tank cleaning personnel must be carefully controlled to ensure continuity. Consideration must be given to holding a further toolbox talk.
- *Tank cleaning.* On completion of the tank cleaning operation, the master must carry out an inspection together with the tank cleaning contractor supervisor to ensure that the tanks have been properly cleaned and lines and pumps are thoroughly flushed. If these parties disagree, an independent surveyor will carry out an inspection. The various commonly accepted tank cleaning standards are shown at *Annex M (Tank Cleaning Standards)*. The tank inspection should confirm that the tanks have been cleaned to the appropriate standard.

9.23 CARRIAGE OF TUBULAR CARGOES

The purpose of this section is to describe the recommended practice for the safe transportation and handling of tubular cargo on offshore supply vessels. In accordance with the current regulations, it is the responsibility of the vessel master to ensure that cargo is properly secured before departure.

- *Tubular cargo* is defined as any type of round objects which are shipped not in separate cargo carriers but using slings to bundle one or more such objects together in a bundle.
- *Pup joints* are short casing or tubing joints used as 'space out' for connecting pipeline sections of a predetermined length.

- *Centralisers* are devices fitted to the outside of the casing/liner to align it to the centre of the bore hole during cementing.

9.23.1 Cargo Requirements

Slinging must be in accordance with regulatory requirements, and properly secured with wire clamps or similar, for example using Welcro bands. Units shorter than 6 metres (19.6 ft) should be transported as a cargo unit. The slinging of tubular cargo must ensure the bundles remain stable during transit. Tubular cargo should preferably be bundled in odd numbers where practicable. Where the cargo consists of standard 9 5/8–13 3/8[2] casings, and fitted with a centraliser, consider having only one tubular in each bundle as it may be difficult to split them on the pipe deck. The slinger must take into consideration the working load limit (WLL) of the slings and the weight of each tubular when slinging the bundles. Certified lifting points fitted on the tubular cargo should be used during the loading of large and heavy dimensions if they cannot be strapped in a prudent manner or handled in certified cargo carriers. Always inspect for loose and/or damaged protectors during each phase prior to the lifting of the cargo.

9.23.2 Preparations before Loading at Base

The vessel must be informed of the tubular cargo well before loading. This information should include the dimensions, weight, length, and quantity of the cargo. Best practice suggests dedicating the most suitable deck area based on destination, which crane will be used, and prevailing weather reports. Hull loads and reduced stability caused by tubular cargo becoming waterlogged must be taken into consideration upon assignment of area. A sufficient safe zone must be established both fore and aft of the dedicated cargo area and must be no less than 1 metre (3.2 ft) in length.

9.23.3 Loading at Base

A representative from the vessel, preferably the officer responsible for loading, must monitor and supervise the loading operation. It is important to ensure bundles are stowed as close together as possible to avoid the risk of shifting cargo during the voyage. When loading large dimensions with one tubular in each bundle, it is best to evaluate whether to fit wedges below each tubular joint to avoid the risk of shifting cargo during transportation or offloading. If wedges are used, these should be nailed to a wooden deck where possible to reduce the risk of shifting. Large dimensions must never be loaded on top of smaller dimensions. When stacking cargo, take into consideration the strength of the deck, as well as the working height for seamen. Two metres (6.5 ft) is normally the maximum stacking height. Vessels must always be loaded in a manner that possible easy securing of remaining cargo onboard in case offloading is interrupted. If possible, tanks and other frame or skid-type cargo units should not be positioned just fore or aft of tubular cargo due to the risk of snagging. Slings on bundles should be extended and laid across the tubular cargo to avoid becoming wedged between bundles. Always determine the appropriate placement in relation to openings and escape routes in cargo rails. Cargo units should not be used as the only barrier to secure tubular cargo on vessel decks. The risk of shifting cargo is normally highest during the voyage to and from the facility. In the event of marginal weather conditions, the risk of shifting tubular cargo must be taken into account when selecting the time of departure, the route taken, and the vessel's speed over ground.

9.24 OFFSHORE LOADING/OFFLOADING OPERATIONS

When preparing to load or offload the vessel, it is imperative to conduct an internal Pre-Job Talk to assess and clarify the following operational details:

- Communications
- Positioning of the vessel
- Distribution of work and roles between the seamen on deck when two pendants and hooks are used
- Operation-specific issues such as the weather, type of tubular cargo, location, any securing arrangements

A Pre-Job Talk between the vessel and the facility crane operator should also take place to clarify the following:

- Communications
- How many bundles will be used for each lift (recommended two bundles maximum)
- Any use of tag lines during offloading to the installation
- Positioning of the vessel as regards vessel movements, reach, and line of sight from the crane
- Operation-specific issues, including risk of snagging

When offloading at the installation, pay special attention during the removal of lashings used throughout the voyage to the oilfield. It is important to use the correct footwear (protective footwear covering the ankles), especially if personnel are required to walk atop tubular cargo. Focus on correct dogging. It is recommended that two eyes are used in each hook depending on the lifting equipment. The deck crew, hook, and cargo must remain within a clear line of sight of the derrick operator. Good radio discipline is essential. Always avoid the use of tag lines if possible. If tag lines must be used, fasten and prepare these before dogging the individual lifts. The risk of snagging on the vessel deck and cargo rails, as well as in potential blind zones, must be accounted for during the positioning of the vessel. In addition to the issues addressed earlier, the following issues are important during loading operations when the vessel is on station. The vessels must be informed of the type, quantity, and weight of the cargo to be returned well before loading starts. The ship's staff must prepare the necessary hawsers as well as chains and pipe support. All tubular cargo to be returned to the vessel should be washed first to avoid incidents arising from slippery tubular cargo on the vessel deck. It is important to use correct footwear (protective footwear covering the ankles) in the event personnel are required to walk atop tubular cargo. Tubular cargo shorter than 6 metres (19.6 ft) should be shipped in baskets. If possible, avoid tubular cargo where the ship's staff must unhook or hook lifting yokes. Tag lines should not be used during the loading of return cargo. The ship's staff must never touch lifts of tubular cargo or walk underneath such lifts before the lift has been landed properly. Slings on bundles must be extended and laid across the tubular cargo to avoid them becoming wedged between bundles.

During the loading of return cargo, always pay special attention to rolling cargo. In connection with large dimensions and if the vessel is rolling, any vessel without automatic sea fastening arrangements (ASFA) or equivalent must use wedges to secure large dimension cargo before unhooking it. It may be useful to have the vessel list somewhat towards the side where the first lifts will be landed. In the event of interrupted offloading or loading at the offshore facility, the vessel must be able to and be given sufficient time to secure the remaining cargo in a proper manner.

9.25 LOADING OF TUBULAR CARGO ONTO PIPELAYING VESSELS

The loading of tubular cargo for pipeline installation projects is normally handled by the pipelaying contractor that has chartered the vessel. Lifting beams are typically used during offloading of this type of tubular cargo, as they are lifted by inserting each end of the tubular cargo into the lifting equipment. In the event of large heights, start loading from the middle to avoid working towards the outer perimeter of the cargo deck, which presents an increased risk of falling overboard. Hull loads and reduced stability calculations resulting from the weight of the tubular cargo, including water inside the tubulars, must be factored in the stability calculations. Perhaps the best method for loading and stowing tubulars on deck is the *Monsvik Method* as it prevents large open spaces between the pipeline bays.

9.26 CARRIAGE OF OIL-CONTAMINATED CARGO

The offshore industry, in conjunction with the Chamber of Shipping and the Marine Safety Forum (MSF) has produced guidance to assist offshore operators in the safe and efficient carriage and offloading of oil contaminated cargoes. During well operations, water-based fluids such as seawater, brine, or water-based mud may become contaminated, commonly with oil-based mud or base oil from oil-based mud. These contaminated products are referred to as wet bulk waste. Because of the presence of oil, these wet bulk wastes cannot be legally discharged to the marine environment. Subsequently, the contaminated fluids are returned to shore for treatment or disposal at dedicated reception facilities. Operations giving rise to such wet bulk wastes include well bore clean-up operations, where oil-based mud is displaced from the well bore to seawater or completion brine; operations where water-based mud becomes contaminated with oil-based mud during displacements; cementing operations with associated spacers; pit cleaning operations; drilling operations where well bore fluids are contaminated with oil-based mud, crude oil, or condensate; other tank cleaning operations where fluid chemical components cannot be discharged because of the *Offshore Chemical Regulations*; and rig floor drains where the fluid is oil contaminated.

Any of the above wet bulk wastes may also be contaminated with hydrogen sulphide (H_2S), typically from sulphate reducing bacteria (SRB). When fluids are severely contaminated and are of small volume, then general industry practice is to transport the waste back to shore in Tote tanks or some other similar type carrying unit. For fluids that are 'lightly' contaminated, general practice is to backload to the mud tanks onboard the offshore supply vessel. It is this latter practice in particular that has raised grave concerns for the following reasons:

(1) It is difficult to accurately describe the chemical make-up of the waste and hence provide a MSDS sheet that adequately describes the material.

(2) Gas testing on offshore supply vessels returning to shore with this type of cargo has found, on a significant number of occasions, high levels of H_2S in the atmosphere above the cargo. Lower Explosive Limit (LEL) tests have also revealed an explosive atmosphere in excess of that which the offshore supply vessel has the capability to safely transport.

(3) The mud tanks on offshore supply vessels are not designed to contain and transport wet bulk cargo with a flashpoint of less than 60°C (140°F). The pump rooms and pumping systems for the discharge of the product tanks are not intrinsically safe. This classification is only found onboard specialist type offshore vessels. The reason for the very high LEL % values that have been recorded is contamination with crude oil and condensates. The bulk mud tanks on standard offshore supply vessels are not designed for this purpose and under *no circumstances* should fluids contaminated with the products mentioned earlier be backloaded to an offshore supply vessel's mud tank.

Recognising the complex nature of oilfield by-products and cargoes, the consortium of offshore operators has developed further procedural guidance for the handling and backloading of wet bulk wastes onto offshore supply vessels. Central to this guidance is the requirement to carry out field tests. These tests can be performed either on the offshore facility or onshore but must be performed by a competent person as determined by the operator. The rate at which fluids are generated during normal offshore operations may preclude sending samples onshore for testing, which can necessitate offshore facility-based testing instead. In either case, the results of the tests must be made available to the vessel master prior to the backloading hose connection taking place. Once the tests have been carried out, no more fluid should be added to the intended cargo. If any further additions are made, a new test is required. The results of these tests should allow the master to establish whether the backload is acceptable for carriage onboard the vessel. Acceptance is based on certain data parameters such as the reported analytical information and the measured physical properties, the known nature of the chemical composition, and any previous cargoes loaded to the vessel's tanks. A generic risk assessment should be carried out and subsequently updated when new, improved, or different information and circumstances become apparent. The offshore facility's personnel should be aware that in certain circumstances the vessel master may require advice from the vessel's onshore technical advisors and that a response from onshore may take time to progress. If there is any doubt regarding the results of the field tests, these should be repeated until confidence is achieved. The backload hose should not be sent to the vessel and connected unless there is agreement between the vessel master and the OIM that the backload is acceptable for carriage.

9.27 COMPOSITION OF THE WET BULK WASTE

The final wet bulk waste may contain components and formulated mixtures that include water (both seawater and potable water), oil-based mud, base oil, water-based mud, well bore clean-up detergents, completion brine (including corrosion inhibitors, biocides, etc.), cement spacers, rig wash, brines containing various salts, and a variety of other toxic and non-toxic substances such as glycol, pipe dope, etc. That being said the major component is normally seawater. The proportions of the other constituents are variable. The wet bulk waste is likely to be heterogeneous in that oil mud will separate to the bottom, base oil to the top, with seawater in between. Vessel motions will not normally be sufficient to mix and stabilise the cargo into a homogeneous form. The components and formulated mixtures may arise from different well bore operations. The volumes of each component are normally known, although the degree of volumetric accuracy is variable, depending on how and where the material is stored on the facility prior to backloading. During discharge to onshore storage tanks and road tankers, the composition of the initial discharge may be different to that discharged later due to separation of components during transportation. This may result in higher concentrations of an individual component being transported in road tankers. For example, oil-based mud typically has the following chemical composition:

Seawater	75%
Mineral oil-based mud	10%
Cement spacer with surfactants	10%
Base oil	5%

This mixture will separate, leaving the base oil on the surface, the seawater below this, and the mineral oil mud at the bottom. The cement spacer will mix with the seawater, although the surfactants will also mix with the base oil and oil mud. During transfer operations from the vessel to road tankers, the initial fluid comprises the heavy oil mud, followed by the lighter seawater, and finally the base oil. In the event of a hose rupture or spillage, all component fluids should be treated as oil contaminated and should be contained, preventing discharge to the sea. Wet bulk waste may contain a significant number of chemicals for which MSDS are required. It is not practicable, however, to develop a description of the wet bulk waste from such an array of documents. Although MSDS will be available for formulated mixtures, there may still be uncertainty in describing the properties of the wet bulk waste. As a precaution, the following tests should be carried out, prior to backloading:

pH	numerical range	0–14
Salinity (chlorides)	mg/l	
Retort	oil content	volume %
	Water content	volume %
	Solids content	volume %
Flashpoint	(closed cup °C/°F)	
Noxious gases	LEL Explosive gases	
	H_2S	
	Oxygen	
Bulk density specific gravity		

As described earlier, tests may be carried out offshore on the facility by trained and competent personnel. Alternatively, samples may be sent onshore for analysis by the waste processor or other competent laboratory. The analysis and treatment should be carried out in a timely fashion on representative samples of each wet bulk waste intended for backloading to the vessel. If backloading is delayed for any reason, such as inclement weather or heavy seas, this should be noted on the analysis form. The pH of the wet bulk waste should be monitored daily. If there is any doubt whatsoever regarding the test results, the test should be repeated until confidence is restored. The results of the tests, along with the analyst's signature and the date completed, should be enclosed with the analysis form and attached to the appropriate Waste Consignment Note, for example, SEPA C note.

In this chapter, we have covered quite a substantial amount of detail regarding the safe carriage of cargoes onboard offshore supply vessels. As we can probably recognise, the task of shipping materials and equipment to offshore facilities is a hazardous and risky business. The type of cargoes carried by offshore supply vessels are often in themselves inherently hazardous.

Add to this arduous nature of the environment these vessels operate in, and the proximity to offshore structures makes cargo operations in the offshore sector by far one of the most dangerous. In the next chapter, we will look at three notoriously difficult cargoes which are collectively referred to as 'special cargoes'. These are timber products, grain, and livestock.

Chapter 10

Principles for the Safe Handling, Stowage, and Carriage of Timber Products, Grain, and Livestock

10.1 INTRODUCTION

In this chapter, we will briefly discuss the main issues around the transportation of three types of 'special cargo': timber products, grains, and livestock. These three types of cargo are notoriously hazardous and can lead to serious incidents if the correct loading and discharging procedures are not followed. Because of their intrinsically hazardous nature, specific regulations pertain to the carriage of these cargoes. Whilst we will not go into the detail of what the regulations specify, by the end of the chapter we should have a clear understanding of the risks and hazards, and the correct procedures to ensure these cargoes are carried safely. To start with, we will begin with timber products.

10.2 TIMBER

Timber is carried on deck because the high stowage factor means that the ship is not down to her marks when the holds are full. The carriage of timber on deck requires special attention for various reasons. One of these is the timber's ability to absorb moisture, thus affecting the vessel's stability. On the other hand, the presence of timber on deck provides for additional buoyancy to the vessel. Because of the several variables that are involved in the carriage of timber deck cargo, the IMO has set out a list of recommendations applicable to vessels greater than or equal to 24 metres (78.7 ft) in length, engaged in the carriage of timber deck cargo. These are set out in the *Code of Safe Practice for Ships Carrying Timber Deck Cargoes*. Timber deck cargoes comprise logs, pit props, or sawn. The stowage factor of timber can range from 27 m^3 (88.5 ft^3)/tonne to 149 m^3 (488.8 ft$^{3)}$)/tonne. When timber is carried on deck, moisture absorption makes the vessel top heavy, thus reducing the stability of the vessel. The chances of capsizing are thus increased. Some allowance is

DOI: 10.1201/9781003354338-11

made for the additional buoyancy provided by the cargo on deck. Some types of timber can absorb up to one third of their weight of water. Apart from the stability criteria, when loading timber on deck, other factors such as the strength of the deck, the possible additional weight involved and shoring and securing arrangements must also be taken into consideration. Timber ships often develop a list if, on their voyage, they experience weather mainly from one side of the vessel, causing the timber to absorb water on one side only from spray.

10.3 STABILITY CRITERIA FOR VESSELS CARRYING TIMBER DECK CARGO

The stability criteria recommended for passenger and cargo ships carrying timber deck cargo are the area under the righting lever curve (GZ curve) should not be less than 3.15 metre-degrees up to 30 degree angle of heel and not less than 5.16 metre-degrees up to 40 degrees or the angle of flooding (Θf) if this angle is less than 40 degrees. Additionally, the area under the righting lever curve (GZ curve) between the angles of heel of 30 degrees and 40 degrees or between 30 degrees and Θf if this angle is less than 40 degrees, should not be less than 1.72 metre-degrees. The righting lever GZ should be at least 0.20 metre at an angle of heel equal to or greater than 30 degrees. The maximum righting arm should occur at an angle of heel preferably exceeding 30 degrees but not less than 25 degrees. The initial metacentric height GM should not be less than 0.15 metre (Figure 10.1).

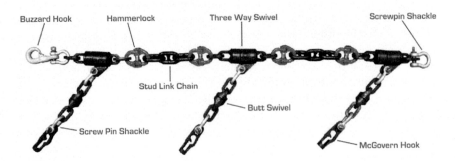

Figure 10.1 Typical timber rigging.

Source: Courtesy of Shaw, New Zealand, email permission.

10.4 PRECAUTIONS NECESSARY WHEN CARRYING TIMBER DECK CARGO

The carriage of timber deck cargo requires several precautions to be taken. The following are some of the main precautions necessary when handling, stowing, and carrying timber deck cargoes (Figure 10.2):

- Openings to be closed and battened down
- Steering arrangements not obstructed
- Cargo compactly stowed and level surface
- Crew's access to be safe
- Strong uprights
- Lashing to be in accordance with regulations
- Quick cargo release mechanism
- Lashing to be regularly inspected
- Vessels to comply with stability criteria
- Water absorption causes excessive stress on lashings
- Entries relashing in logbook
- Prudent ship-handling
- Height of cargo on deck limitations
- Provisions for guard rails or lines
- Protection of personnel

Figure 10.2 The correct timber stowage on deck.

Source: Email permission.

Timber itself is not a cargo particularly liable to damage. It does not require as much care as other cargoes. However, it can be the source of problems for other cargoes, the vessel, and its crew. It is from this perspective that the carriage of timber deck cargo must be seen and attended to. One can never be too cautious when dealing with cargo liable to hugely affect the stability of the vessel.

Extract from the Code of Safe Practice for Timber Deck Cargo

SHIPS CARRYING TIMBER DECK CARGOES

1. *SCOPE*
   1.1 The provisions given hereunder are recommended for all ships of 24 metres or more in length engaged in the carriage of timber deck cargoes.
   1.2 Administrations are invited to adopt these recommendations unless they are satisfied that operating experience justifies departures therefrom.
   1.3 Nothing in these recommendations shall preclude the application of Regulation 44 of the 1966 Load Line. Convention as far as ships with timber loadlines are concerned or any national requirements.
   1.4 For the purpose of these recommendations 'Timber Deck Cargo' or 'Cargo' means a cargo of timber carried on an uncovered part of a freeboard or superstructure deck, which should include logs, sawn timber whether loose or packaged.

2. *STOWAGE – GENERAL*
   2.1 Openings in the weather deck on which cargo is stowed should be securely closed and battened down. Ventilators and air pipes should be effectively protected from damage by cargo and the check valves in air pipes should be examined to ascertain that they or similar devices are effective against the entry of sea water.
   2.2 The cargo should be compactly stowed and should be chocked, as necessary, for this purpose; it should not interfere in any way with the navigation and necessary work of the ship and should be stowed as level as practicable. Safety equipment, devices for remote operation of valves and sounding pipes should be clearly marked and left accessible.
   2.3 Steering gear components should be efficiently protected from damage and the arrangements made for steering in the event of a breakdown in the main steering gear should not be obstructed by deck cargo.
   2.4 Safe and satisfactory access to the crew's quarters, machinery spaces and all other forward and after parts used in the necessary

working of the ship should be provided at all times. Cargo in the vicinity of the openings which give access to such parts should be so stowed that the openings can be properly closed and secured against the entry of water.

2.5 Uprights should be of adequate strength considering the nature of the timber and the breadth of the ship; the strength of the uprights should not exceed the strength of the bulwark and the spacing should be suitable for the length and character of timber carried but should not exceed 3 metres. Strong angles or metal sockets or equally efficient means should be provided for securing the uprights. Where suitable, permanent ship's structure may be used as uprights.

2.5.1 The uprights should be of such height as to extend above the outboard top edge of the cargo.

2.5.2 They should preferably be fitted with a locking pin or other arrangement to retain the upright in its housing.

2.5.3 They may be secured by a metal bracket attaching the upright to the top of the ship's bulwark, or a similar arrangement.

2.5.4 As far as is practicable, the stowage should be such that the cargo throughout its full height is in solid contact with each upright.

2.5.5 Each port and starboard pair of uprights should be linked by athwartship lashings, set up taut joining each pair as near to the top level of the cargo as possible to give additional strength to these uprights. The lashings should be in accordance with the recommendations of Section 4.

3. *LASHINGS – GENERAL*

3.1 Timber deck cargo should be efficiently secured throughout its length by independent overall lashings. The spacing between adjacent lashings should be in accordance with subsections 5.2 and 5.3 such that each lashing should pass over the cargo and be shackled to eye plates. Eye plates for these lashings should be efficiently attached to the sheer strake or to the deck stringer plate. The distance from an end bulkhead of a superstructure to the first eye plate should be not be more than 2 metres. Eye plates and lashings should be provided 0.6 metre and 1.5 metres from the ends of timber deck cargoes where there is no bulkhead. Stanchions and brackets or other such points of insufficient strength should not be used for the securing of lashings.

3.2 Stretching devices or similar devices for lashing shall be either a turnbuckle or of a type that produces tightening by a lever action having a proven mechanical arrangement whereby tightening to the required tension and subsequent adjustments can be rapidly effected; each specified device should be so placed in a lashing that it can be safely and efficiently operated when required.

3.3 When such devices are of a portable type, a sufficient number should be carried onboard.

3.4 A device capable of quick cargo release, if fitted, should be so designed that it cannot be accidentally released or activated.

3.5 When other devices are substituted for splicing to form an eye in wire rope, they should be sufficient to equal the strength of the splice.

3.6 Lashings should be inspected as required.

4. *LASHINGS – TESTING AND CERTIFICATION*

4.1 All testing, marking, and certification of the chains should conform with national regulations.

4.2 In addition to the requirements stated, a visual examination at intervals not exceeding 12 months is recommended.

4.3 The lashings should be capable of withstanding an ultimate load of not less than 133 Kn (13,600 kp). They should be fitted with slip hooks and tumbuckle, which should be accessible at all times. Wire rope lashings should have a short length of long link chain to permit the length of lashings to be regulated.

4.4 After testing there should be no treatment applied to chain which would invalidate its test certificate, for example, galvanizing heat treatment.

10.4.1 Timber and Cargo Securing

Timber cargoes are unique, in that they are robust and they are buoyant. With these factors in mind, consideration has been given by Statutory Authorities and Classification Societies to recognising the cargo's contribution to a ship's reserve buoyancy when loaded with timber that is properly secured. Therefore, there are two distinct aspects to timber cargoes, the first being the contribution to reserve buoyancy by the cargo and the assignment of an alternative maximum loaded draught and secondly the securing arrangements.

1 Openings to be closed and battened down.
2 Steering arrangements not obstructed.
3 Cargo compactly stowed and level surface.
4 Crew's access to be safe.
5 Strong uprights.
6 Lashing to be in accordance to regulations.
7 Quick cargo release mechanism.
8 Lashing to be regularly inspected.
9 Vessels to comply with stability criteria.

10 Water absorption causes excessive stress on lashings.
11 Entries in respect to relashing to be made in the logbook.
12 Prudent ship-handling.
13 Height of cargo on deck – limitations.
14 Provisions for guard rails or lines.
15 Protection of personnel.

10.4.2 Load Line

Vessels that carry timber can be assigned an alternative set of draught marks based on the contribution of the cargo to the ship's reserve buoyancy, that is the vessel's seaworthiness. However, the cargo must be stowed and secured in a very specific manner. The following extract from the *International Convention on Load Lines, 1966*, as corrected by the *Procès-Verbal of Rectification* dated 30 January 1969 and the *Procès-Verbal of Rectification* dated 5 May 1969.

10.5 REGULATION 44 – STOWAGE

10.5.1 General

(1) Openings in the weather deck over which cargo is stowed shall be securely closed and battened down. The ventilators shall be efficiently protected.

(2) Timber deck cargo shall extend over at least the entire available length which is the total length of the well or wells between superstructures. Where there is no limiting superstructure at the after end, the timber shall extend at least to the after end of the aftermost hatchway. The timber shall be stowed as solidly as possible to at least the standard height of the superstructure.

(3) On a ship within a seasonal winter zone in winter, the height of the deck cargo above the weather deck shall not exceed one-third of the extreme breadth of the ship.

(4) The timber deck cargo shall be compactly stowed, lashed and secured. It shall not interfere in any way with the navigation and necessary work of the ship.

10.5.2 Uprights

(1) Uprights, when required by the nature of the timber, shall be of adequate strength considering the breadth of the ship; the spacing shall be suitable for the length and character of timber carried, but

shall not exceed 3 metres (9.8 feet). Strong angles or metal sockets or equally efficient means shall be provided for securing the uprights.

10.5.3 Lashings

(2) Timber deck cargo shall be efficiently secured throughout its length by independent over-all lashings spaced not more than 3 metres (9.8 feet) apart. Eye plates for these lashings shall be efficiently attached to the sheer strake or to the deck stringer plate at intervals of not more than 3 metres (9.8 feet). The distance from an end bulkhead of a superstructure to the first eye plate shall be not more than 2 metres (6.6 feet). Eye plates and lashings shall be provided 0.6 metres (23½ inches) and 1.5 metres (4.9 feet) from the ends of timber deck cargoes where there is no bulkhead.

(3) Lashings shall be not less than 19 millimetres (¾ inch) close link chain or flexible wire rope of equivalent strength, fitted with sliphooks and turnbuckles, which shall be accessible at all times. Wire rope lashings shall have a short length of long link chain to permit the length of lashings to be regulated.

(4) When timber is in lengths less than 3.6 metres (11.8 feet) the spacing of the lashings shall be reduced or other suitable provisions made to suit the length of timber.

(5) All fittings required for securing the lashings shall be of strength corresponding to the strength of the lashings.

10.5.4 Stability

(6) Provision shall be made for a safe margin of stability at all stages of the voyage, regard being given to additions of weight, such as those due to absorption of water and icing and to losses of weight such as those due to consumption of fuel and stores.

10.5.5 Protection of Crew, Access to Machinery Spaces, etc.

(7) In addition to the requirements of Regulation 25 (5) of this Annex guard rails or lifelines spaced not more than 330 millimetres (13 inches) apart vertically shall be provided on each side of the deck cargo to a height of at least 1 metre (39½ inches) above the cargo.

10.5.6 Steering Arrangements

(8) Steering arrangements shall be effectively protected from damage by cargo and, as far as practicable, shall be accessible. Efficient provision shall be made for steering in the event of a breakdown in the main steering arrangements.

10.6 REGULATION 45 – COMPUTATION FOR FREEBOARD

(1) The minimum summer freeboards shall be computed in accordance with Regulations 27 (5), 27 (6), 27 (11), 28, 29, 30, 31, 32, 37 and 38, except that Regulation 37 is modified by substituting the following percentages for those given in Regulation 37 (Table 10.1):

Table 10.1 Percentage of Deduction for All Types of Superstructure

	Total Effective Length of Superstructure										
	0	0.1L	0.2L	0.3L	0.4L	0.5L	0.6L	0.7L	0.8L	0.9L	1.0L
Percentage of deduction for all types of superstructure	20	31	42	53	64	70	76	82	88	94	100

Percentages at intermediate lengths of superstructures shall be obtained by linear interpolation.

(1) The Winter Timber Freeboard shall be obtained by adding to the Summer Timber Freeboard one thirty-sixth of the moulded summer timber draught.
(2) The Winter North Atlantic Timber Freeboard shall be the same as the Winter North Atlantic Freeboard prescribed in Regulation 40 (6).
(3) The Tropical Timber Freeboard shall be obtained by deducting from the Summer Timber Freeboard one forty-eighth of the moulded summer timber draught.
(4) The Fresh Water Timber Freeboard shall be computed in accordance with Regulation 40 (7) based on the summer timber load waterline.

What does this all mean? In the example of a small ship (over 24.0 m but under 85.0 m in length) where the superstructure is 30% of the vessel's length, a timber deck cargo, properly loaded and secured, would result in a reduction in the ship's freeboard. That is the cargo is considered to contribute to the ship's superstructure. If the ship had a continuous superstructure, it would be permitted a reduction of 350 millimetres to its minimum freeboard (see Regulation 37). From Table 10.1, it can be seen that the reduction would comprise 53% of 350 millimetres or 185.5 millimetres. That is more cargo can be carried because the ship's maximum loaded draught has been increased.

10.6.1 Securing

Marine Orders Part 42 lays down the legislation governing the Securing of Cargo onboard. As from January 1998, all vessels must have a Cargo Securing Manual made specifically for that vessel. The Cargo Securing

Manual is made following recommendations from the Code of Safe Practice for Cargo Stowage and Securing (the CSS Code) and the *Guidelines for the Preparation of the Cargo Securing Manual.* The Cargo Securing Manual is required to contain details of the securing arrangements available together with the equipment, the amount, and location of such and the test that they have been subject to.

10.6.2 Summary

- Timber cargoes contribute to a reduction in a vessels freeboard if properly loaded and secured.
- Timber cargoes are potentially hazardous in that they can cause a substantial reduction in a vessel's stability by uneven absorption of water.
- There are three sets of regulations that affect timber cargo operations directly and coxswain and master need to be aware of.

10.7 GRAIN

Grain cargo is another one of these 'special' cargoes which requires special attention from the carrier and their agents. Various regulations are in place to make the carriage of grain as safe as possible. The *International Code for the Safe Carriage of Grain in Bulk,* more commonly referred to as the *IMO Grain rules,* forms the backbone of the regulatory framework for the carriage of grain products by sea. In the UK, the *Merchant Shipping (Carriage of Cargoes) Regulations 1999,* in conjunction with the *IMO Grain Code,* govern the carriage of free-flowing grain. That is to say, the rules do not apply to ships carrying grain in bags or in containers. The *IMO Grain Rules* defines grain as 'wheat, maize (corn), rye, oats, barley, rice, pulses, seeds and processed forms thereof, whose behaviour is similar to grain in its natural state'. In accordance with the IMO Grain Rules, each country enforces its own interpretation of how vessels must declare their intention to load grain products to the relevant Port State Authority. In Australia, for example, this is done through the submission of form MO 33/1, which is provided in Figure 10.3. Once the declaration has been submitted and approved by the Port State Authority, a document of authorisation is issued to the grain-carrying vessel certifying the vessel complies with the *IMO Grain Rules.* The document of authorisation is issued by the Flag State and forms part of the Grain Loading Manual that is maintained onboard. Vessels without a document of authorisation may be allowed to carry grain provided the vessel can demonstrate that she complies with the main provisions as specified in the Code.

MO 33

Australian Government
Australian Maritime Safety Authority

NOTIFICATION OF LOADING, OR SAILING AFTER PARTIAL DISCHARGE, OF BULK GRAIN

Marine Order 33 Cargo and Cargo Handling - Grain

GENERAL

Chapter VI of the SOLAS 1974 Convention, as amended, and Australian legislation (Marine Order 33 [Cargo and Cargo Handling – Grain] require that ships intending to carry grain cargoes in bulk from Australian ports may be requested to demonstrate compliance with the International Grain Code.

SOLAS 1974 requires the cargo shipper to provide the Master or his representative with appropriate information on the cargo. Beyond this it is the Master's responsibility to ensure the proper stowage of the cargo in accordance with Marine Order 33.

INSTRUCTIONS TO MASTERS

This notification is required to be submitted to AMSA at least 72 hours prior to the vessels proposed commencement of loading.

In the case of intending to sail after partial discharge, a completed notification must be submitted to AMSA at least 24 hours prior to the anticipated time of sailing.

The master or Agent is required to submit the form to the nearest AMSA office of the port at which grain is to be loaded or partially discharged (see over).

A grain stability calculation using form AMSA 226, the 'Calculation of Stability for Ships Carrying Bulk Grain', must be completed by the ship and retained for inspection by an AMSA surveyor during an onboard inspection.

A separate form is required to be submitted for each Australian port. The master may lodge all the notifications to the relevant office prior to the first port of call or may lodge them individually to each appropriate office.

Strict adherence to the layout of this form is not necessary as long as the information required by it can be provided by alternate means.

A new form is required to be submitted to AMSA if there is any significant change in the loading plan.
The master must ensure their declaration is signed and dated before sending the notification to AMSA. Do not enter anything in the box marked 'To be Completed by the Surveyor'.

The surveyor will advise the master whether an inspection is required or not. This advice will be sent to the person who provided the notification and may be by e-mail, either with the notification attached (with the surveyor's advice completed), or by e-mail that includes the same advice in the body of the message.

NOTES

AMSA applies the following provisions when assessing compliance with the Code:

1. AMSA does not accept "partly filled" compartments untrimmed, even if data for these is approved by the flag State Administration, as they are not provided for in the Code.
2. Some Australian grain loading terminals lack the facility to adequately trim the ends of filled compartments and masters must check the facilities at their load ports if they consider they need to trim the ends of any compartments in order to meet the required stability criteria.
3. AMSA cannot accept a compartment as being "filled" if the average ullage at the coaming exceeds the minimum required to accommodate the structure of hatch covers or 100mm, whichever is greater.
4. Untrimmed moments may only be used for filled compartments with the ends untrimmed.
5. In partly filled compartments AMSA accepts grain surfaces in which the height between the highest peaks and the lowest troughs in the compartment is not more than 1.0m as being "level" within the meaning of the Code and therefore trimmed to an acceptable level.
6. It is the responsibility of the Master to ensure that the cargo is trimmed as required by the Code - AMSA will not determine the method by which this is achieved.

AMSA225 (4/18)

Figure 10.3 Notice of intention to load bulk grain.

LODGEMENT OF GRAIN FORMS

For the purposes of Sections 13 and 15 of Marine Order 33 (Cargo and Cargo Handling – Grain), the Manager Ship Inspection and Registration has approved the following methods for lodgement:

QUEENSLAND

Brisbane
Mail:	AMSA – Operations North
	PO Box 10790
	Adelaide Street
	Brisbane QLD 4000
Fax:	07 3001 6801
Email:	bneoperations@amsa.gov.au

Gladstone
Mail:	AMSA – Operations North
	PO Box 297
	GLADSTONE QLD 4680
Fax:	07 4972 3841
Email:	gltoperations@amsa.gov.au

Mackay
Mail:	AMSA – Operations North
	PO Box 42
	Mackay Post Office
	Sydney Street
	Mackay QLD 4740
Fax:	07 4957 8450
Email:	mkyoperations@amsa.gov.au

VICTORIA and TASMANIA

Melbourne;
Geelong;
Portland; and
Tasmanian Ports
Mail:	AMSA – Operations South
	PO Box 16001
	Collins Street West
	MELBOURNE VIC 8007
Fax:	03 8612 6003
Email:	mlboperations@amsa.gov.au

SOUTH AUSTRALIA

Port Adelaide;
Port Lincoln;
Port Pirie;
Port Giles;
Wallaroo;
Ardrossan; and
Thevenard
Mail:	AMSA – Operations South
	PO Box 3245
	Port Adelaide, SA 5015
Fax:	08 8447 3855
Email:	ADLOperations@amsa.gov.au

NEW SOUTH WALES

Port Kembla
Mail:	AMSA – Operations East
	PO Box K976
	HAYMARKET NSW 1240
Fax:	02 8918 1390
Email:	sydoperations@amsa.gov.au
or	
Mail:	AMSA – Operations East
	PO BOX 102
	Port Kembla NSW 2505
Fax:	02 4274 7806
Email:	sydoperations@amsa.gov.au

Newcastle
Mail:	AMSA – Operations East
	PO Box 86
	CARRINGTON NSW 2294
Fax:	02 4961 2694
Email:	nsoperations@amsa.gov.au

WESTERN AUSTRALIA

Geraldton;
Kwinana;
Bunbury;
Esperance; and
Albany
Mail:	AMSA – Operations West
	PO Box 1332
	FREMANTLE WA 6959
Fax:	08 9430 2121
Email:	freoperations@amsa.gov.au

AMSA225 (4/18)

Figure 10.3 (Continued)

Australian Government
Australian Maritime Safety Authority

LOADING, OR SAILING AFTER PARTIAL DISCHARGE, OF BULK GRAIN   MO 33

NOTICE OF INTENTION TO LOAD BULK GRAIN

Marine Order 33 Cargo and Cargo Handling   Grain

This form is required to be submitted to AMSA – see Instructions to Masters and Notes

SHIP DETAILS

Name of ship	IMO number	Type of ship ☐ Bulk Carrier ☐ Tween Decker ☐ Other (specify):		
Flag	Gross tonnage	Summer deadweight	Summer draught	Year keel laid
Agent (for cargo and contact details)				

Approving authority for Document of Authorisation	Date of approval	
If applicable for the intended loading/voyage and the Flag State Administration has issued a sheltered water exemption as permitted by A 5 of the Grain Code: Area covered by the exemption:	Date of issue	Date of expiry

CARGO DETAILS (A separate form is required for each port and is to represent the total cargo on board on departure from that port)

Total number of holds: Approved Stability booklet provided for: ☐ untrimmed ends ☐ trimmed ends ☐ both

Hold No.	Type of grain/cargo	Stowage factor	Tonnes	% Full (See Note 2 on Page 1)	Grain trimmed or untrimmed (T/U)	Stability calculated using trimmed/ untrimmed moments (T/U)
Example	BARLEY	1.37	6168	100	U	U

Throughout the voyage the highest actual heeling moments will be   and the maximum allowable heeling moments will be
Maximum angle of heel   (12° maximum) *(To be completed if vessel's grain loading booklet does not include a table of allowable heeling moments or where the actual KG and Displacement fall outside the parameters of the table).*

IN-TRANSIT FUMIGATION

Is in-transit fumigation to be carried out on this cargo?
No ☐ Yes ☐ ➔ If yes, provide name of fumigator:
Note: The Fumigator is required to notify AMSA of the intention to fumigate. They may use <u>AMSA 82</u> for this purpose.

An AMSA surveyor may request verification of the above loading and stability at any time prior to the ships departure.

Any vessel loading or discharging grain at an Australian port may be subject to inspection by AMSA at any time to ensure compliance with the Code.

Master's Certification

This is to certify that:

1. The intended loading is as per the above and the vessel's stability is prepared in accordance with the requirements of the vessel's Grain Loading Booklet and the International Grain Code, if the loading changes, AMSA will be advised;
2. The vessel will comply with the requirements of Parts 7, 8 or 9, as applicable, of the International Grain Code at all stages of the voyage;
3. Form AMSA 226, 'Calculation of Stability for Ships Carrying Bulk Grain', has been completed, and is ready for presentation on board to AMSA if requested;
4. Bulk grain will be stowed as per the requirements of Part 10 of the International Grain Code; and
5. During loading, on departure, and throughout the voyage the vessel's bending moments and shear forces will not exceed the allowable limits;
6. If fumigation is required, MSC.1/Circ.1264 is to be followed, in particular the following information have been reviewed and complied with:
   - Evidence of Flag State acceptance of arrangements
   - Master agrees to proposed arrangements
   - Evidence that the Fumigator is appropriately licenced
   - PPE is provided and adequate to fumigation
   - Ship spaces will be monitored in accordance with IMO guidelines through the entire period
   - Other requirements, including those of the Port Authority and flag State, will be complied with.
7. Sailing Draught: F: A: M: ;
8. Next port of call: ;
9. If required the Ship will be ready for inspection at: Port: Berth:
10. Intended date and time of loading Date: Time:

Master's signature:.. Name (printed):   Date:

AMSA225 (4/18)

Figure 10.3 (Continued)

TO THE MASTER – To be Completed by the Surveyor

1. An inspection by an AMSA Surveyor   is / is not* required before the ship commences loading.
2. If an inspection is required, additional evidence is to be available to demonstrate compliance with the International Grain Code (see AMSA 226).

*delete as required.

Surveyor: *Signature* .... *Name* .... Date   /   /

Figure 10.3 (Continued)

10.8 STOWAGE OF BULK GRAIN

When loading and discharging grain cargo, all necessary and reasonable trimming must be performed to level all free grain surfaces and to mini-mise the effect of grain shifting. If there is no other cargo above a cargo of bulk grain, then the hatch covers must be secured in an approved man-ner. When bulk grain is stowed on a closed tween deck which is not grain tight, the covers should be made grain-tight by taping the joints. Unless already allowed for in the vessel's stability assessment, the surface of the bulk grain in a partly filled compartment should be secured to prevent grain shift by over-stowing, strapping, or lashing, and by rigging shifting boards. Grain must be kept dry and with good ventilation. Poor ventila-tion can lead to moisture developing in the hold, which in turn can cause fermentation. The stowage factor for grain varies according to the type of grain and whether it is shipped in bulk or in bags. In any case, the cargo must always be stowed away from sources of heat and moisture. The pri-mary risks associated with the carriage of grain cargo include germina-tion,1 infestation by rodents and insects, dust explosion, grain swelling, and grain shifting (Figure 10.4).

10.8.1 Germination

It is important that grain is kept dry as it may germinate. No loading or dis-charging should be done in wet conditions. Ship sweat may also cause grain to germinate as will any contact with bilge water.

10.8.2 Infestation

Typically, fumigation is carried out to prevent infestation. This requires the compartment to be thoroughly cleaned with any traces of previous cargo removed. Special attention is required if the previous cargo was liable to produce vermin or other insects. A grain loading permit will only be issued after an independent surveyor has inspected the compart-ment, during which the surveyor will pay special attention to the risk of infestation.

Figure 10.4 Ship loading wheat grain.

Source: John Fiddes, CC BY-SA 2.0.

10.8.3 Dust Explosion

This is likely to occur from the dust produced when grain is loaded. In the right proportion and in suspension in the atmosphere, dust can explode when a source of ignition is applied. Smoking is strictly prohibited when grain is being handled and the atmosphere is dust laden.

10.8.4 Grain Swelling

This occurs when grain absorbs moisture. The moisture causes the grain to swell and this in turn exerts additional pressure on the structures of the compartment. Distortion of the compartment can cause severe structural damage to the vessel.

10.8.5 Grain Shifting (in a Filled Compartment)

If the compartment is full and trimmed, 'saucering' is one method of securing the cargo. This is done by trimming the cargo in a slight saucer shape and placing a tarpaulin over the top of the indentation. Bagged grain is then tightly stowed over the tarpaulin. The hatchway must also be loaded with bagged grain. Alternatively, bundling is another option available in a full and trimmed compartment. In this case, the cargo is trimmed using the saucer method, with other materials laid on top. Between the bulk grain and the material wire lashings running athwartship are put in place. Cargo is then loaded on top of the material and the lashings are drawn up tight which creates a large bundle above the bulk grain.

10.8.6 Grain Shifting (in a Partly Filled Compartment)

There are several potential methods of managing a partly filled compartment. The first involves the over-stowing method. Here, the surface of the grain is levelled and covered with separation cloths or bearers. Bagged grain is then tightly stowed on top of the material or platform. Cargoes other than bagged grain may be used provided they exert the same pressure. The second method employs straps and lashings. Here, the cargo is trimmed in a slightly crown shape. Prior to the completion of loading, wires are secured to the sides of the hold. On completion, the cargo is covered with a separation cloth on top which runs on athwartship and fore and aft bearers. Steel wires are made to run from one side of the hold to the other, on top of the bearers and tightened with turnbuckles. During the voyage, as the grain settles, the lashings must be tightened to maintain sufficient torque. The third method involves securing the cargo with wire mesh. This method is like the strapping method except that wire mesh (the same type as those used to reinforced concrete) is laid atop of the separation cloth.

In some countries, Port State Authorities stipulate that shifting boards must be erected whenever grain is carried. Shifting boards are wooden centre lines that are fitted in the hold and extend to one third of the depth of the hold from the top and cannot be less than 2.45 metres (8.03 ft) deep. The shifting boards prevent the grain from shifting athwartship after settling post loading (Figure 10.5).

10.9 VESSEL STABILITY

In addition to the document of authorisation mentioned previously, vessels engaged in the carriage of grain must also have a set of approved grain stability data, which illustrates various loading conditions. These important documents are reviewed by the independent surveyor prior to giving their permission to load grain. Essentially, any vessel intending to carry grain must demonstrate to the authorities that the vessel complies fully with the stability criteria. Thus, the vessel master must complete the GA Form showing their stability calculation for the intended voyage. Vessel stability calculations are performed in accordance with Part B of the IMO Grain Code. Part B of the Code deals with the

Figure 10.5 Ship loading rapeseed.
Source: Herve Cozanet, CC BY-SA 3.0.

calculation of assumed heeling moments and general assumptions. The criteria for vessels carrying grain cargoes are set out as follows:

(1) The angle of heel due to the shift of grain shall not be greater than 12 degrees, or in the case of ships constructed on or after 1 January 1994, the angle at which the deck edge is immersed (Θf, whichever is the lesser.
(2) In the statical stability diagram, the net or residual area between the heeling arm curve and the righting arm curve up to the angle of heel of maximum difference between the ordinates of the two curves, or 40 degrees, the angle of flooding (Θf, whichever is the least, shall in all conditions of loading be not less than 4.3 metre-degrees.
(3) The initial metacentric height, after correction for the free surface effects of liquids in tanks, shall be not less than 0.30 metres (0.98 ft).

The grain trade is a major part of sea transportation. The safe and efficient carriage of grain cargoes requires special care and attention. Unless the dangers associated with their carriage and the ways in which they can be damaged are understood, ship owners risk huge cargo claims. And these claims could easily be avoided.

10.10 LIVESTOCK

Australia is one of the world's main exporters of livestock together with New Zealand and Brazil. As such, the AMSA and the Australian national legislature have developed extensive laws and regulations governing the carriage and welfare of livestock by sea. Any vessel involved in the shipment of livestock from or to Australia is required to comply with these regulations. Subsequently, the carriage of livestock is another area where some expertise is required. Strict national legislation may be temporarily put into place to reduce the risk of disease if it is found that such a risk exists. Previous examples have included bovine spongiform encephalopathy (BSE disease in cows) and Avian influenza (bird flu) in chickens and other fauna. These regulations may even apply to animals transiting a particular region. Other than the regulatory requirements, there is an intrinsic need for the humane and hygienic treatment and conditions for animals transported by land and sea. At its most basic, the animals, which range from birds to reptiles and mammals, must be looked after in a humane and compassionate way (Figure 10.6).

Figure 10.6 Typical livestock carrier – MV *Livestock Express.*

Source: Christian Ferrer, CC BY-SA 4.0.

The major concern in the carriage of livestock is the welfare of the animals being carried. Every effort should be made to ensure that they are well looked after, and travel in a decent environment. They should be provided with sufficient food and clean water, be able to move, and have adequate lighting. They should have clean air and be protected from any source of danger, inclement weather, and extreme conditions. Livestock can be carried in containers, in cages, or simply in pens on decks. The stowage position must ensure that they can be accessed easily (if veterinary attention is required or to remove any carcasses) and be protected from the extremes of heat, cold, rain, wind, and sea spray. In Australia, the regulations governing the carriage of livestock are stated in Marine Order Part 43 (Figure 10.7).

Australian Government
Department of Agriculture
and Water Resources

Imported goods

Biosecurity risk treatment guide
Compliance Division
Version 6.0

Figure 10.7 AQIS quarantine treatments and procedures.

Source: AMSA/AQSA.

This applies to all ships on which it is intended to take livestock at any port in Australia and to Australian ships whenever they are carrying livestock. It does not apply, however, to a vessel that has loaded livestock at a port outside Australia and which is to be discharged in Australia.

10.11 LOADING PROCEDURES (AUSTRALIA)

Before any livestock can be loaded onboard a vessel, a notice of intention to load livestock must be given to the surveyor in charge. The surveyor will then check the vessel to ensure it is suitable to carry livestock. Towards the end of the loading procedure (usually 3 hours), the vessel master must advise the surveyor on the estimated time of completion. Livestock must not be loaded without the authorisation of a Government-approved veterinary. Furthermore, all certificates required by the Australian Commonwealth Department of Primary Industries and Energy must be issued and reviewed by the surveyor. Vessels which have permanent fittings erected for the carriage of livestock are issued with an Australian Certificate of Livestock. A Certificate of Approval evidences any item of equipment that must be approved under Marine Order Part 43.

10.12 STABILITY CRITERIA FOR VESSELS CARRYING LIVESTOCK

The stability criteria that apply to vessels carrying livestock may be summarised as follows: the area under the righting lever curve must not be less than 3.15 metre-degrees (0.055 metre-radians) up to 30 degree angle of heel and not less than 5.16 metre-degrees (0.09 metre radians) up to 40 degree angle of heel, or the angle of flooding if this angle is less than 40 degrees. The area under the righting lever curve between the angles of heel of 30 degrees and 40 degrees, or between 30 degrees and the angle of flooding (Θf) if this angle is less than 40 degrees, must not be less than 1.72 metre-degrees (0.03 metre-radians). The righting lever must not be less than 0.20 metre at an angle of heel equal to or greater than 30 degrees. The maximum righting lever must occur at an angle of heel not less than 25 degrees. The initial metacentric height must not be less than 0.15 metre. The area under the righting curve, up to 40 degrees or the angle of flooding, whichever is less, in excess of the area under the heeling lever curve to the same limiting angle, must not be less than 1.03 metre-degrees (0.018 metre-radians) plus 20% of the area of the righting lever curve to the same limiting angle. The angle of heel due to wind must not be more than 10 degrees. In calculating the stability of the ship, the use of fuel oil, freshwater, and fodder, the movement of ballast and the build-up of waste material must be accounted for (Figure 10.8).

Figure 10.8 Typical livestock carrier – *MV Hannoud.*

Source: Ken Hodge, CC BY-SA 2.0

10.13 RESTRICTIONS ON THE CARRIAGE OF LIVESTOCK

Livestock must not be carried if the livestock or any livestock fitting or equipment obstructs access to any accommodation space necessary for the safe running of the ship; interferes with life-saving appliances or firefighting apparatus; interferes with the sounding of tanks and bilges; interferes with the operation of closing appliances; interferes with the operation of freeing ports; interferes with the lighting or ventilation of other parts of the ship; and/or interferes with the proper navigation of the ship.

10.14 PROVISION OF VENTILATION, LIGHTING, AND DRAINAGE

When livestock is carried within enclosed spaces, there must be a mechanical way to change the air within that space. A surveyor may test that system prior to loading. In some circumstances, a mechanical ventilation system might not be necessary, especially if there is sufficient headroom for air circulation. The mechanical system must have a primary and a secondary source of power to be approved. There must be lighting providing a level of illumination of not less than 20 lux in all areas and passageways where livestock are carried. Provisions must be made for the effective draining of fluids from each pen in which livestock are carried. Contaminated water must not be intentionally discharged from the ship whilst within the 12 nautical mile (13.8 mi; 22 km) limit.

10.15 PROVISION OF FIREFIGHTING APPLIANCES

There must be sufficient fire hydrants so that at least two jets of water can be directed simultaneously to any part of a space where livestock are located. These must always be accessible and be operated by any member of the ship's staff.

In this chapter, we have looked at three types of special cargoes: timber products, grains, and livestock. Hopefully, it should be clear that these types of cargoes require special handling and stowage. Because of their intrinsically hazardous or sensitive nature, special regulations must be complied with to ensure vessel and cargo safety. In the second part of this book, we will turn our attention to the use of cargo-handling equipment and the methods for securing and stowing cargo onboard.

NOTE

1 The sprouting of a seed, spore, or other reproductive body, usually after a period of dormancy. The absorption of water, the passage of time, chilling, warming, oxygen availability, and light exposure may all contribute in initiating the germination process.

Part 2

Safe Handling, Stowage, and Securing of Cargo

LEARNING OUTCOME 2

On completion of this section, you should be able to safely handle, stow, and secure different types of cargoes, including IMDG cargoes.

ASSESSMENT CRITERIA

2.1 The hold and tank preparation procedures for the reception of various cargoes are explained.

2.2 The methods of de-odourising and fumigating holds are outlined.

2.3 Methods of handling and stowage of cargo using machinery are outlined.

2.4 The importance of securing the cargo is discussed.

2.5 The function and types of dunnage used in stowing and securing cargo are outlined.

2.6 Recommended procedures of securing a range of cargoes are outlined.

2.7 Calculations involving weights, capacities, stowage factors, and load densities are performed.

2.8 Terminology used in the carriage of dangerous goods is defined.

2.9 The classification and markings of dangerous goods are explained.

2.10 The precautions that should be taken during the loading, carriage, and discharge of dangerous goods are stated.

2.11 The content of IMDG Code and its supplement is outlined.

2.12 The requirements stated in the Marine Orders with regard to the following are outlined:
- Safety precautions and proceedings during cargo operations.
- Notices that should be forwarded to the authorities.
- Proposed loading plan and final stowage plan.
- Surveys carried out onboard.

2.13 Importance of cargo care is discussed.

2.14 The causes of cargo damage are listed, and their effect and preventative measures are explained.

DOI: 10.1201/9781003354338-12

2.15 The methods of temperature control of refer cargoes are outlined.

2.16 The causes, effects, and prevention of sweat are explained.

2.17 The causes, effects, and prevention of contamination of cargo are outlined.

2.18 The principles of cargo ventilation are explained.

2.19 The cargo ventilation methods and systems are outlined.

2.20 The function of dunnage with respect to cargo care is explained.

2.21 Cargo damage survey procedures are outlined.

Chapter 11

Requirements Relating to the Use of Cargo-handling Equipment and Cargo Stowage

11.1 INTRODUCTION

Cargo can be handled in various ways with various tools. These are commonly known as cargo gear and are very important for smooth cargo operations. They accelerate the handling rate and reduce the risk of damage caused by improper handling. These items of equipment must be well-maintained and checked each time before use. A common standard must exist to determine the suitability of the gear. In Australia, this is the role of Marine Order Part 32. Marine Order Part 32 states all the checks, tests, and examinations that must be carried out on cargo gear before and after use. It also states who is responsible for the maintenance and performance of tests. In this chapter, only some of the most important sections are covered and explained. This in no way diminishes the importance of the other sections. Thus, reading the relevant sections of Marine Order Part 32 will complement the information provided in this chapter. It is important to recognise that Marine Order Part 32 applies to the loading or unloading of any ship at a port in Australia or in an external territory of Australia; the loading or unloading at any port of a ship to which Part II of the Navigation Act applies, that is, an Australian registered ship, a ship engaged on the coasting trade, or a ship owned or operated by an Australian company with Australian crew; and the loading or unloading of an offshore industry mobile unit.

Note: Whilst this chapter refers specifically to the Australian regulations pertaining to the use of cargo-handling equipment and cargo gears, each country will have their own specific regulations. For the UK, this will be the (COSWP) and the Provision and Use of Work Equipment Regulations (PUWER). Readers are strongly encouraged to research which regulations are relevant to them.

11.2 KEY TERMS AND DEFINITIONS

To begin with, it is worth spending a few moments explaining some of the key terms that relate to the use of cargo gear. *Cargo gear* refers to any article

DOI: 10.1201/9781003354338-13

of equipment which is designed for use with a crane or derrick in loading or unloading cargo that is:

(1) Not riveted, welded or otherwise permanently attached to the crane or derrick; or
(2) Designed to be detachable from the crane or derrick and includes any wire rope, fibre rope, sling, net, clamp, grab, loose gear, magnetic lifting device, vacuum lifting device, patent-handling system, or self-unloading system but does not include transport equipment or packaging.

Materials-handling equipment refers to any article, or an integrated assembly of articles, designed to convey or for use in conveying cargo, and includes cargo gear, a crane, derrick, cargo lift, side loading platform, mechanical loading appliance, and/or mechanical stowing appliance. *Proof load*, in relation to materials-handling equipment, means the proof load for that equipment which is determined in accordance with Marine Order Part 32. *Safe working load* (SWL) means, in relation to an article of materials-handling equipment, the load that a responsible person considers is the maximum load that may be imposed on that article to allow an adequate margin of safety in the normal operation of that article.

11.3 LOADING AND UNLOADING OPERATIONS

Loading and unloading must not be carried out on a ship unless it is in full compliance with Marine Order Part 32. Power-operated hatch covers can only be operated by a crew member, or someone authorised by the officer in charge of the loading or unloading operation. The ship's side doors, bow doors or stern doors can only be operated by a crew member or a person who has been specifically authorised by the vessel master.

11.3.1 Safe Working Loads

An article of materials-handling equipment must not be used in loading and/or unloading a ship unless a responsible person, having regard to the design, strength, material of construction, and proposed use of the article, has determined the safe working load of the article, and thereafter marked the safe working load and associated information on the article, in accordance with Marine Order Part 32. Any article of materials-handling equipment must not be used to handle a load exceeding the safe working load of that equipment except in accordance with Appendix 6 of Marine Order Part 32. When calculating the load on derricks or cranes, the mass of any loose gear, spreader, equalising beam, and such like attachments must be accounted for.

11.3.2 Testing, Examination, Inspection, and Certification of Wire Rope

Wire rope may only be used if (1) a responsible person has issued a certificate in respect of the rope in accordance with Marine Order 32/4 Appendix 23; (2) a competent person has inspected the rope, externally and, as far as practical, internally, and found that the rope is not worn, corroded, or otherwise defective to a degree that renders it unfit for the proposed use; (3) the rope is free from knots and kinks; (4) the rope complies with the structural requirements specified in Appendix 15 of Marine Order 32; and (5) evidence, based on prototype testing, that any terminal or end fitting on the rope complies with Appendix 5 of Marine Order 32 is recorded in the appropriate register of materials-handling equipment. A wire rope must not be used if the total number of broken wires visible in a length of the rope is equal to ten times its diameter and exceeds 5% of the number of wires constituting the rope.

11.3.3 Register of Materials-handling Equipment and Certificates of Testing

The vessel must carry, onboard, a register of the materials-handling equipment for use by the ship. This register must contain particulars of all tests, examinations, inspections, heat treatments, and any maintenance, repair, or replacement of materials-handling equipment to which Marine Order Part 32 applies. All certificates of testing and other relevant certificates must be kept with, or near at hand to, the register of materials-handling equipment.

11.3.4 Provision of Protective Fencing

Loading and unloading must not be carried out unless there is in place protective fencing to prevent any person from falling. These fences must be in accordance with Appendix 1 of Marine Order Part 32. This is particularly important where cargo is loaded or unloaded in the lower holds, and where personnel are working in the tween decks.

11.3.5 Provision of Lighting

Loading and unloading must not be carried out unless there is adequate lighting in place. An illumination of at least 20-lux is considered adequate.

11.3.6 Provision of Safe Atmospheres

Loading and unloading must not be carried out if the cargo space is liable to contamination by harmful concentrations of dust or toxic vapours, or in which there is a likely deficiency of oxygen.

11.3.7 Provision of Fire Extinguishing Systems and Equipment

Internal combustion engines (ICE) or electric motors must not be used in a cargo space in connection with loading and/or unloading unless there are adequate fire extinguishing systems or equipment suitable for extinguishing fuel-based and electrical fires.

11.3.8 Safe Use of Materials-handling Equipment

A load, other than, for example, a spreader or cargo-lifting beam, must not be left suspended from, or supported by, a derrick, or crane unless during the time it is suspended or supported, a qualified person is at the control position of the equipment. Shackles and other similar devices must be effectively secured against accidental dislodgment or release. A load must not be dragged by means of a runner leading from a derrick or crane if there is a risk that the SWL of any component of the derrick, crane, or associated cargo gear would be exceeded. A person must not be hoisted or lowered in the course of cargo operations by means of a crane or derrick other than in a personnel cradle.

11.4 REQUIREMENTS FOR DERRICKS

A derrick for use in the loading or unloading of cargo must be marked with its safe working load for each operating condition, as well as the lowest angle to the horizontal at which the derrick may safely be used. The marking must include the letters SWL followed by numerals representing the SWL and letters representing the unit used. If there is more than one operating condition, then an oblique stroke separating the units of mass for each condition must be clearly visible. For example:

SWL × t or SWL x/y t

If the derrick is to be used in conjunction with a purchase rig, then the letters 'SWL' must be followed by the letter 'u' in brackets and the numerals representing the SWL. For example:

SWL (u) × t or SWL (u) x/y t

11.5 REQUIREMENTS FOR CARGO GEARS

A wire rope must not be used in loading or unloading unless in the case of a rope for use other than as a guy pendant, a preventer guy, a guy, a stay or a net or sling:

- It contains at least 114 constituent wires.
- Any fibre materials in its construction are of strand or rope core only.
- In the case of a runner or purchase, it comprises one continuous length without joins.
- Any thimble or loop splice fitted to the rope complies with Appendix 22 of Marine Order 32.
- It is free from knots or kinks.

11.6 SIGNALLING

To avoid confusion and accidents, signalling must only be done in accordance with Marine Order Part 32, Appendix 18.

In addition to Marine Order Part 32, publications such as the *BSMA 48/76* and *Code of Safe Practice for Seamen* also provide useful guidance regarding the safe use of materials-handling equipment when cargo work is involved. In Australia, AMSA has published the *Code of Safe Working Practice for Australian Seafarers*. This is available free of charge from AMSA or can be downloaded from AMSA's website at www.amsa.gov.au. It should be borne in mind, though, that any cargo operations must be done in strict accordance with the relevant regulations as it is a legal requirement to do so. For Australian vessels, the ship's cargo officer is strongly encouraged to familiarise themselves with the contents of Marine Order Part 32 so as to carry out their duty in a fully diligent manner.

11.7 HANDLING AND STOWAGE OF DIFFERENT CARGOES USING MACHINERY

As stated previously, cargo must be handled, stowed, and carried properly. This entails using the proper equipment. To carry out this task successfully, different types of equipment are used to handle different types of cargo. The main reasons for doing so are as follows:

- The cargo is not damaged.
- Cargo operations take less time.
- Less labour is used thus, it is more economical for the shipper.
- Standard procedures are adopted in various parts of the world.

11.7.1 Handling of Bulk Cargo

Bulk cargoes will be loaded or handled by grabs, chutes, or conveyor belt. This will depend on the port facilities, the type of bulk cargo, and the vessels cargo-handling gear. Figure 11.1 shows the various methods by which bulk cargo can be handled.

Figure 11.1 MVV Due at Kwinana bulk terminal.

Source: A Calistemon, CC BY-SA 4.0.

11.7.2 Handling of Bulk Liquid

Bulk liquid will invariably be handled by hoses with pump rooms located on deck. Manifolds and crossover valves will be used to direct the cargo to the desired compartment. In some cases, some part of the circuit might be dedicated to only one type of cargo to avoid contamination (Figures 11.2–11.4).

11.7.3 Handling of Containers

The handling of containers nowadays is done in a very fast and efficient way. Large ports will have dedicated container terminals with gantries and container stackers, while small ports might use the ship's cranes and container spreader to handle the boxes. It must be borne in mind that the handling of containers requires some expertise from the ship officer. A vessel with a certain amount of list might not be able to load any containers if loading is being done by ship or shore gantry. Planning the location of each container is critical. Often, container stowage plan is drawn up by shore terminal and difficult to fully check (Figures 11.5–11.8).

11.7.4 Handling of Cargo on RORO Vessels

RORO vessels do not normally carry gears, that is, they do not have cargo-handling gear. Instead, the cargo is driven on and off the vessel through big openings such as ramps or through hatch openings (Figures 11.9–11.11).

Figure 11.2 MV Sifnos at the Port of Fremantle.

Source: A Bahnfrend, CC BY-SA 4.0.

Figure 11.3 Bulk unloader system for large bulk vessels.

Source: Ashley Dace, Magellan Maritime Press Ltd.

Figure 11.4 Tilbury grain port and silos.

Source: Ashley Dace, Magellan Maritime Press Ltd.

Figure 11.5 MV Banaster at CBH Grain Jetty.

Source: A. Calistemon, CC BY-SA 4.0.

Figure 11.6 Grain loading at KSK terminal.

Source: Ivan Studenov, CC BY-SA 4.0.

Figure 11.7 Typical container terminal.

Source: Pexels.

Figure 11.8 Containers being handled by the ship's gears – *MV Maersk Vyborg*.

Source: Author's own.

Figure 11.9 Handling of containers ashore using a straddle carrier.

Source: Author's own.

Figure 11.10 Krupp double-joint deck cranes.

Source: US Navy, Open Government Licence.

Figure 11.11 Manoeuvring a load into the cargo compartment on a RORO vessel.

Source: Author's own.

11.7.5 Handling Heavy Weights

Heavy lift carriers or general cargo ships equipped with heavy lift gear normally handle heavy and oversized cargoes. When working heavy lift cargoes, special ballasting arrangements need to be put in place to maximise stability. Also, adequate dunnage needs to be laid to spread the weight of the heavy lift cargo (Figures 11.12–11.14).

11.8 HOLD AND TANK PREPARATION PROCEDURES FOR THE RECEPTION OF VARIOUS CARGOES

This subject has been dealt with in Part 1 of this book, but as a reminder, the preparations typically involve preventing damage to the following:

- Cargo through contamination, water absorption, sweat, movement, and fire
- Ship due to movement of the cargo, reaction with the cargo, spillage, and pollution
- Environment by way of pollution, oil, dust, ballast, and bilge water

Figure 11.12 Manoeuvring a load into the cargo compartment on a RORO vessel.

Source: Author's own.

Figure 11.13 Discharging cargo from a RORO vessel – *MV Maersk Willow*.

Source: Author's own.

Figure 11.14 Small container ship with dual derricks – *MV Malik Arctica*.

Source: Andrzej Otrebski, CC BY-SA 4.0.

11.9 FUMIGATION AND DE-ODOURISATION OF CARGO HOLDS

When preparing a cargo compartment for loading, it is sometimes necessary to fumigate and de-odourise the compartment. Though the two processes are often associated with each other, they have very different purposes. Fumigation, for example, is carried out when vermin are believed to exist and present a danger to the cargo and vessel. It is associated with moths, bugs, and rats that thrive on some types of commonly carried cargo. The three main reasons why fumigation may be necessary are as follows:

(1) To destroy residual infestation from previous cargoes
(2) To fumigate a particular cargo, for example, malt, corn (maize)
(3) To comply with quarantine requirements

De-odourising, on the other hand, is necessary when residues of previous cargoes may taint sensitive cargo such as fruits. De-odourisation, as the name suggests, involves the removal of odours which have the potential to adversely affect any incoming cargo.

11.9.1 Fumigation

Some cargoes are particularly attractive to vermin. Bales, wood, animal skins, and grains are all good examples of excellent breeding grounds for vermin. Moreover, insects might be present in the compartment even long after a particular cargo has been discharged. If left unchecked, insects will proliferate and attack the cargo within the compartment. They can also be responsible for spreading diseases. Rats are carriers of many types of bacteria, viruses, and other pathogens. Also, they feed on any edible goods being carried. In the interest of health as well as the preservation of cargo, fumigation is at times essential. There are two primary ways of fumigating the compartment. Specialist companies who are trained and experienced in the safety procedures for such operations normally carry out the fumigation process. There are, however, certain fumigants which are safe for use by inexperienced operators. Fumigation is most often needed when contact insecticides cannot be used in an efficient way. Methyl bromide is a commonly used fumigant in container fumigation.

The container should have warning labels prominently posted around it and be fenced off from all unauthorised approach. Phosphene may be introduced in the form of pellets which emit a gas when in contact with air. The residue of the pellets must be removed before cargo-handling personnel are permitted to enter the enclosed space. Both methyl bromide and phosphene are dangerous gases, and as such every effort should be made to warn personnel of the presence of these chemicals. The dangers of methyl bromide are outlined in Marine Notice 12/1972, a copy of which is provided in Figures 11.15 and 11.16.

Maritime and Coastguard Agency

MARINE GUIDANCE NOTE

MGN 86 (M)

RECOMMENDATIONS ON THE SAFE USE OF PESTICIDES IN SHIPS

Notice to operators, shipowners, charterers, masters, agents, port and harbour authorities, shippers, container and vehicle packers, cargo terminal operators, fumigators, fumigant and pesticide manufacturers and all persons responsible for the unloading of freight containers

(This notice takes immediate effect)

Summary

The purpose of this guidance note is to advise on:

(1) the importance of safe and proper procedures when pesticides are used on board ships;

(2) the appropriate application of the IMO Recommendations on the Safe Use of Pesticides in Ships both to cargo and to cargo spaces; and

(3) to indicate the likely application of other related requirements or guidance which would be applicable to the use of, handling or transport of pesticides.

1. In accordance with the Merchant Shipping (Carriage of Cargoes) Regulations 1997[1], where pesticides are used in the cargo spaces of ships prior to, during, or following a voyage, the IMO publication "RECOMMENDATIONS ON THE SAFE USE OF PESTICIDES IN SHIPS" (1996 Edition, IMO267E), available from IMO, Publications Section, 4 Albert Embankment, London SE1 VSR where relevant thereto shall be complied with. The contents of this publication are also incorporated into the Supplement to the International Maritime Dangerous Goods (IMDG) Code. The use of pesticides includes the fumigation of cargo spaces and of cargo, in port, or in-transit, and any part of the ship so affected by their use, as contained in the Recommendations.

2. The Maritime and Coastguard Agency, in conjunction with the Ministry of Agriculture, [1]Fisheries and Food, the Scottish Office, Agriculture, Environment and Fisheries Department and the Health and Safety Executive, considers that it is essential that adequate precautions are taken by all those responsible for the commissioning of pest control on board ships. MCA strongly recommends observance of all the provisions contained in the Recommendations, but the necessity for the master and crew to cooperate with shore-based fumigation personnel in compliance with other safety requirements should be recognised. In the United Kingdom the Health and Safety Executive (HSE) is the relevant shore-based authority.

[1] S.I. 1997/No.19 as amended

Figure 11.15 Marine Guidance Note MGN86 (M) recommendations on the safe use of pesticides in ships.

3. Mandatory requirements cover the conditions for preparation and carriage of cargo transport units under fumigation. These are classified as Class 9 dangerous goods with the proper shipping name "CARGO TRANSPORT UNIT UNDER FUMIGATION" in the International Maritime Dangerous Goods (IMDG) Code, ("cargo transport unit" being any freight container or vehicle shipped under fumigation). The Merchant Shipping (Dangerous Goods and Marine Pollutants) Regulations 1997 require compliance with the IMDG Code for packaged dangerous goods for aspects such as declaration, stowage, segregation, marking, labelling and the display of a fumigation warning sign.

4. Pesticides, when not in use and carried as cargo may also be subject to the Regulations specified in paragraph 3 above and the IMDG Code.

5. The Recommendations regarding fumigation practice were written as a result of consultations with experts on pest control, pesticide safety and ship operation.

6. In one case, failure to comply with the recommended procedures caused a number of people to be hospitalised after exposure to phosphine gas generated in a cargo of grain fumigated with Aluminium Phosphide during the sea passage. The fumigant tablets were not fully decomposed and, hence, the fumigation process was not fully completed before the vessel arrived at the discharge port. There have also been a number of other incidents involving containerised cargoes arriving under fumigation at United Kingdom ports with no accompanying documentation on the ship or at the port of discharge regarding the nature of the cargo.

7. Merchant Shipping Notice MSN 1718 should be referred to for the statutory requirements on the safe use of pesticides (including fumigants), in cargo spaces on board ships when they are loaded or intended to be loaded with cargo. Although extensively referred to in these requirements, the scope and application of the IMO Recommendations is generally wider, providing, for example, guidance on the disinfestation of food stores, galleys and crew and passenger accommodation.

8. As pesticides are hazardous substances their handling and application and exposure to them are subject to regulations affecting the health and safety of workers at work. For further information other related documents should be referred to, e.g. Chapter 27 of the "Code of Safe Working Practices for Merchant Seamen"[2] and the Health and Safety Executive Approved Code of Practice (Control of Substances Hazardous to Health in Fumigation Operations"[3].

---

[2] Available from The Stationery Office
ISBN 0 11 5518363

[3] HSE Books ACOP L86 and also Guidance note on Fumigation CS22.

MSASD
Maritime and Coastguard Agency
Spring Place
Southampton SO15 1EG
August 1998

Tel: 01703 329 184
Fax: 01703 329 204

File reference: MS 116_31_028

DETR
ENVIRONMENT
TRANSPORT
REGIONS

An executive agency of the Department of the Environment, Transport and the Regions

2

Figure 11.15 (Continued)

MARINE GUIDANCE NOTE

Maritime &
Coastguard
Agency

MGN 497 (M+F)

Dangerous Goods – including Chemicals and other Materials – Storage and Use on Board Ship.

Notice to all Ship Owners, Ship Operators and Managers, Masters and Officers of Merchant Ships, Skippers, Owners & Operators of Small Commercial Code Vessels;, Owners, Operators and managers of Fishing Vessels, Agents, Charterers, Training Providers, Inspectors of Cargoes, Port Authorities, Terminal Operators and others involved in the storage and use of dangerous goods, chemicals and materials onboard ship.

Summary
- This Marine Guidance Note highlights the importance of the correct storage and stowage of packaged quantities of dangerous goods including chemicals, and other materials that are not cargo by virtue of their being in use and/or stored ready for use on board and includes the requirement to carry out a risk assessment in accordance with the ship's Safety Management System.

1. Introduction

1.1 This guidance has been issued following recent incidents and requests for clarification from industry concerning the correct storage of chemicals which are intended for use on that vessel or transferred to another vessel for subsequent use.

It is aimed to address situations where chemicals, materials or goods are in use or stored ready for use whether or not they;

fall within the definition for ships' stores,
are taken out of their prescribed containment systems or stowage locations
and either used on board or transferred for use to an associated daughter craft, workboat or tender.

Where there are particular risks as indicated by the supplier's safety data sheets and/or if they are classified as dangerous goods under the International Maritime Dangerous Goods (IMDG) Code as may be indicated by the goods' original packaging, labelling, documentation etc. The taking on board, storage or direct handling of unfamiliar and

1

Figure 11.16 Marine Guidance Notice MGN 497 (M+F) dangerous goods – including chemicals and other materials – storage and use onboard ship.

suspect goods should be avoided without first consulting the suppliers safety data sheets or obtaining further advice.

1.2 An assessment should be carried out to decide whether goods are physically safe and environmentally suitable for carriage, storage, handling and/or use.

2. The Regulations

2.1 The IMO definition of ships' stores (MSC.1/Circ.1216) is as follows: *Ships stores means materials which are on board a ship for the upkeep, maintenance, safety, operation or navigation of the ship (except for fuel and compressed air used for the ship's primary propulsion machinery or fixed auxiliary equipment) or for the safety or comfort of the ship's passengers or crew. Materials intended for use in commercial operations by a ship are not considered as ships' stores (eg materials used for diving, surveying and salvage operations such as IMDG Code classified dangerous goods (eg Class 1 – Explosives and the other eight classes of dangerous goods).*

2.2 Dangerous goods, not meeting the IMO definition of ships' stores that are carried on board, are subject to the provisions of the IMDG Code. The vessel is therefore required to comply with the relevant provisions of SOLAS and the IMDG Code. In addition, when the dangerous goods are transferred to another vessel (eg a daughter craft), the vessel shall comply either with the requirements of SOLAS or MGN 280, section 30 and be issued with a Document of Compliance for the carriage of dangerous goods.   Further guidance regarding hazardous substances can be found in the Code of Safe Working Practice, Chapter 27.

Storage of Dangerous Goods, Chemicals and Materials

2.3 The importance of the proper storage of Dangerous Goods, Chemicals and Materials on board ship should not be underestimated. If the dangerous goods are subject to the provisions of the IMDG Code, the stowage provisions of chapter 7.1 of the IMDG Code apply. However, a wide variety of chemicals (materials), not subject to the provisions of the IMDG Code, which are commonly used in the marine industry, can react violently together should the packaging become damaged or involved in a fire. Their safe storage should be subject to a risk assessment. The following points should be considered:

2.3.1 **Storage areas** – are designated and controlled areas for portable machinery and equipment containing chemicals, materials, waste, flammable substances e.g. foam plastics, flammable liquids and gases such as propane and hazardous substances e.g. pesticides and timber treatment chemicals; such areas should be arranged so that in the event of a spillage or leakage the substance concerned is contained locally and does not react violently with any nearby substances or materials.

2.3.2 **Accommodation** – storage areas should not be located in or close to accommodation areas;

2.3.3 **Access routes** – corridors and other walkways should not be used as storage areas. Do not store materials where they obstruct access routes or where they could interfere with emergency escape routes;

2.3.4 **Segregation** – store incompatible materials in separate areas; flammable materials will usually need to be stored away from other materials and protected from accidental ignition;

2.3.5 **Safe stowage/Storage at height** – all stores should be securely stowed, if materials are stored at height (e.g. on shelving) make sure necessary guard rails are in place

- 2 -

Figure 11.16 (Continued)

to stop items falling and, if the storage area if fitted with a fixed sprinkler system, ensure that the maximum stowage height limit is adhered to at all times.

2.3.6 **Tidiness** - keep all storage areas tidy, whether in designated stores areas or at a workstation on board the ship; and

2.3.7 **Stock control** - plan deliveries to keep the amount of hazardous materials on board to a minimum, taking into account the operating pattern of the vessel.

Use of dangerous goods and chemicals

2.4 When IMDG Code regulated packaged goods are opened or removed from their prescribed containment system or stowage area whilst on board, the IMDG Code provisions no longer apply.

2.5 Under such circumstances, it is the responsibility of the master/skipper or the company representative under the General Duties provisions of the Merchant Shipping and Fishing Vessels (Health and Safety at Work) Regulations 1997, as amended, to carry out and document a risk assessment to ensure that an equivalent level of safety is maintained when the dangerous goods are transferred to another vessel or are removed from their packaging or containment system. The Merchant Shipping and Fishing Vessels (Health and Safety at Work) (Chemical Agents) Regulations and Marine Guidance Note MGN 409(M+F) may also be relevant. The principles of removing or reducing risks, where possible by replacing with less hazardous substances, minimising risks by reducing exposure and provision of information, training and protection should all be considered. When applicable, the risk assessments should be carried out in accordance with the procedure contained in the ships' Safety Management System (SMS) and a copy kept at the company's offices and on board.

2.6 The risk assessment should address such issues as the physical and environmental characteristics, usage, handling, exposure, stowage, personal protective equipment, and emergency procedures, and the subsequent stowage of opened dangerous goods packaging. From this, safe handling procedures should be developed and where appropriate incorporated into the permit to work system. Companies should use their expertise in the use and handling of dangerous goods and make use of material safety data sheets; where a safety data sheet is used as part of a risk assessment, it is recommended that a copy of the sheet should be kept on board the vessel for easy reference. Emergency responses to spillage of dangerous goods are contained in the IMO Medical First Aid Guide and the IMO Emergency Procedures for Ships Carrying Dangerous Goods (EmS). When appropriate, professional legal and technical advice should be referred to in the assessment of risks and in the preparation of procedures to mitigate the risk.

2.7 Crew should be familiar and competent in the routine and emergency procedures for handling the goods, and any equipment required for carrying these out is readily available at all times.

Figure 11.16 (Continued)

More Information

Environmental Policy Branch
Maritime and Coastguard Agency
Bay 2/08
Spring Place
105 Commercial Road
Southampton
SO15 1EG

Tel :	+44 (0) 23 8032 9141.
Fax :	+44 (0) 23 8032 9104
e-mail:	dangerous.goods@mcga.gov.uk.

General Inquiries: infoline@mcga.gov.uk

MCA Website Address: www.dft.gov.uk/mca

File Ref: MS 116/032/0020

Published: August 2013
Please note that all addresses and
telephone numbers are correct at time of publishing

© Crown Copyright 2013

Safer Lives, Safer Ships, Cleaner Seas

- 4 -

Figure 11.16 (Continued)

Reference may also be made to the AQIS Manual on Quarantine Treatments, Aspects and Procedures, with specific reference to the section on fumigation guidelines. A short extract of which is included here.

11.9.2 Part A – General Information on All Quarantine Treatments

(1) *The structure and use of this manual.* This document provides the standards, and guidance on those standards, for treatment providers wishing to meet Australian quarantine requirements. The document has a modular format. Treatment providers will only need the 'parts' of the manual relevant to the treatment(s) they are performing. For example, a fumigation provider performing methyl bromide fumigation will only need parts A, B and the relevant appendices. Part A of the booklet is relevant and refers to all AQIS quarantine treatments.

It contains general information on quarantine treatments and the responsibilities of all parties involved in those treatments. Each of the subsequent parts are specific to individual AQIS acceptable treatments. For example, Part B is only relevant to methyl bromide fumigation. The information relating to these treatments is detailed. It will provide background to the standards and AQIS requirements. These parts of the manual contain within them the AQIS standards for specific treatments. AQIS intends to add further treatments as they are established. The following flow chart outlines the structure of the AQIS Quarantine Treatments – and Procedures. For example, methyl bromide fumigators would only need Part A, its direct appendices, and all of Part B, and its appendices (i.e. all the shaded boxes) (Figure 11.17).

(2) *Use of contact insecticides.* This process usually involves ship's personnel and can be carried out in two ways:

    (a) Contact insecticides can come in the form of smoke which is discharged into the air as fine particles. It is not effective where there are holes and hard to access places. it is usually done in holds prior to loading of grain; and

    (b) Contact insecticides can be sprayed on particular surfaces. the spray is more effective than the smoke in places where there are holes and cavities since the operator can aim at these holes. It is more time consuming and will require more labour to cover a large surface area.

Figure 11.17 Fumigation of ship's hold – *MV Elefheria.*

Precautions to be taken when fumigating include wearing protective clothing, gloves, respirators and eye protection gear; not removing clothes, gloves, etc. while applying insecticides, even under hot weather conditions; avoiding excessive application and run-off on surfaces; avoiding contamination of foodstuffs; in the event of contact, wash with plenty of water and seek medical advice; full agreement between ship and shore staff must be reached; correct stowage, appropriate ventilation, proper notices and labels must be in place; restricted access for all unauthorised persons; and, smoking may be dangerous in some sprayed atmospheres as some fumigants decompose when heated (Figure 11.18).

Before fumigation begins, post warnings at all entrances. Ascertain that there are no potential sources of leakage. Check that the engine room, ship's accommodation, and working spaces are free of toxic vapours. Check that adjacent spaces are vapour free. Cargo spaces that are sealed for fumigation must never be entered until permission is granted by the master. Prior to entering the space, always monitor the gas content of the space using a gas detector. Once the space is found to be safe, warning signs must be removed. If white pellets are seen with some cargo, these should not be approached until the risk has been fully assessed. If there is any doubt about the safety of the compartment, it should only be entered by personnel donning self-contained breathing apparatus (SCBA).

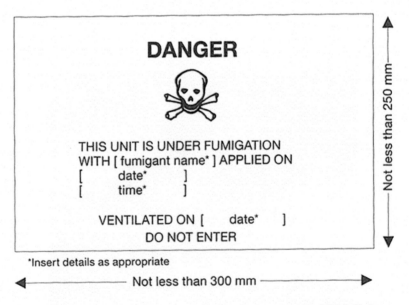

Figure 11.18 Example of fumigation warning sign as demonstrated in IMDG Code.

Source: Author's own.

11.9.3 De-odourisation

In the case of de-odourising, normally the cargo compartment and any adjacent space such as the bilges or trunk are cleaned thoroughly first. If possible, these spaces should be washed with fresh water and non-toxic detergents. Once dry, a de-odourising agent is lightly sprayed into the areas where there are strong odours. Areas in and around the bilges are notorious for harbouring unpleasant odours and smells. Removing the taint reduces the risk of mould growing on the bulkheads and frames. Sometimes new dunnage may have a strong smell. In these instances, it may be necessary to de-odourise the dunnage before loading the cargo. Alternatively, using clean dunnage that has already been used previously is sometimes specified by the shipper to avoid this problem. To avoid concentrations of de-odourising agent, it must be applied only as a fine spray.

In summary, then, in this chapter, we have examined some of the key requirements relating to the use of cargo-handling equipment, the handling and stowage of cargo using machinery, and hold and tank preparations, including fumigation and de-odourisation. In the next chapter, we will discuss the importance of cargo securing and cargo care.

Chapter 12

Importance of Cargo Securing and Cargo Care

12.1 INTRODUCTION

The importance of securing any cargo loaded onboard a vessel for carriage by sea cannot be underestimated, as failing to do so can render the ship unseaworthy. The rules regarding the safe and effective securing of cargo are contained in the *Hague-Visby Rules*. In essence, inadequate lashing of containers and rolling stock on RORO vessels has frequently been found as the cause of cargo shifting, causing the vessel to list and ultimately capsize. As a general rule, once cargo is stowed in position, lashing and securing and tomming must be carried out to prevent even the slightest movement of the load. All lashings must be set up tight; wooden tomming should be secured in such a way that it cannot be dislodged by the ship's vibrations, workings, and movement. As a rule of thumb, a total lashing on one side of a load should have a combined breaking strain of at least 1.5 times the total weight of the load to be restrained. Any tomming and chocking should be in addition to this.

12.2 MAIN OBJECTIVES OF CARGO SECURING

The main aim of securing cargo is to prevent its loss or damage, and/or causing damage to the ship or to other cargo. Furthermore, the loss of the vessel and of lives can result if lashing is inadequate or poorly executed. Exercising a lack of diligence when securing cargo renders the vessel unseaworthy, thus constituting a breach of contract between the shipper and shipowner. To increase the seafarer's awareness on the importance of lashing, the IMO has published a booklet entitled *Guidelines for the Preparation of the Cargo Securing Manual*. Also published by the IMO is the *CSS Code*. In addition to the above publications, in Australia, Marine Order Part 42 lays down the legislation governing the Securing of Cargo onboard. As of January 1998, all vessels must have a Cargo Securing Manual which is specific to that vessel. The *Cargo Securing*

DOI: 10.1201/9781003354338-14

Manual must be produced and carried onboard in accordance with the recommendations laid out in the *CSS Code* and the *Guidelines for the Preparation of the Cargo Securing Manual*. The ship's *Cargo Securing Manual* must contain details of the securing arrangements available onboard together with the equipment, the amount and location of such equipment, and any tests and maintenance procedures that equipment is subject to.

12.3 SECURING OF CARGOES

12.3.1 Securing General Cargoes

The securing of general cargo is very particular, as there is no one single method; each load should be considered in accordance with the particulars of that specific load. The vessel's movement in a sea way must be understood and these external forces counteracted by means of lashing. The physical characteristics of the cargo will inevitably affect the way it is secured, and so this must also be factored in.

- *Securing of palletised cargo:* Cargo on pallets can be secured in many ways such as by strapping, with nets, glue, or shrink wrap. However, the pallets must be properly secured once loaded. Remember, pallets are conducive to block stow.
- *Securing of steel products:* Steel plates and bars which are highly liable to shift can be very difficult to manage depending on their stowage positions. They must as far as possible be levelled and over stowed with other cargoes. Sometimes it might be better to stow them in bins, formed by the bulkhead, the ship's sides, by way of shifting boards. Planking, toms, and shores should be properly secured where used. Depending on the layout of the hold, securing can be optimised by use of block stowage. In some steel trades, bolsters – half-height containers – are used for transporting some steel products, making securing and stowing easier. The stow inside the bolster must be checked, however, to ensure that no movement of the product is possible. It has been known for the lower layer of bolsters to be welded to the ship's tank to prevent lateral movement. When stowing slabs and plate, aim for a solid block stow to avoid the toppling of individual stacks (Figure 12.1).

12.3.2 Securing of Coils

Coils are usually stowed athwartships in regular tiers with the major axes in the fore and aft line with the bottom tier choked off. Particular attention must be given to cargo in the forward compartment of the ship where the effects of heavy pitching are most pronounced. Individual coils in the top

Figure 12.1 Securing of steel plates using the ship's side.

Source: Author's own.

tier of the stow are normally secured by driving wedges between the adjacent coils on either side and fore and aft. Over-stowing a coil cargo with wire rods, bales, or other cargo is an option. In some cases, a floor of dunnage might be necessary (Figure 12.2).

12.3.3 Securing of Logs and Timber Products

Logs stowed fore and aft are normally bundled with wire ropes, while pipes when stowed in holds are normally secured with wire rope and wooden chocks while lying on a cradle base. The securing of timber deck cargo is done according to the *Code of Safe Practice for Ships Carrying Timber Deck Cargoes, 2011* (TDC Code) and with the use of quick release mechanisms (Figure 12.3).

12.3.4 Securing of Cargo on RORO Ships

The securing of cargo onboard RORO ships must be in accordance with an approved system. Securing points and appropriate trestles, etc. should be used to bypass the springing system of vehicles. When selecting the lashing

Figure 12.2 Recommended securing arrangements for coils.

Source: Author's own.

Figure 12.3 Recommended securing arrangements for logs.

Source: Author's own.

equipment to secure cargo on RORO vessels, the following factors must be considered:

(1) The ship's characteristics and motion when underway
(2) The size, weight, and centre of gravity of the cargo onboard
(3) The types of deck surfaces
(4) The position of the cargo load and its supports (wheels, jacks, etc.)
(5) The number, position, and angle of lashing points

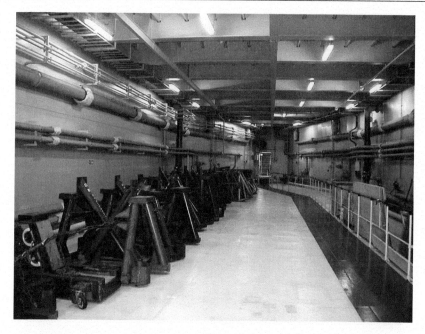

Figure 12.4 Securing arrangement for trailers on a RORO ship.

Source: Author's own.

Normally, the lashing material will comprise chains, wires, and fibre ropes with levers used for tensioning. Turnbuckles also form a major part of the equipment used to lash rolling cargoes. Figure 12.4 shows the securing arrangement of a trailer. The towing vehicle's brake should be applied and locked, lashing then connected to specially fitted lugs on vehicle.

Sometimes chokes and jacks are used in addition to the above. Securing of trailers should encompass the securing of the load on the trailer and the securing of both trailer and load together. Wheel-based cargoes must be provided with adequate and clearly marked securing points. The brakes of such loads must be set prior to securing. The wheels must be blocked to prevent shifting. Securing of vehicles must be in accordance with an approved system, making full use of trestles, pedestals, deck securing points, etc. Securing of containers on RORO ships must be done with the use of locating cones and securing pins. Inspection of securing the cargo on the trailer is necessary before acceptance of load (Figure 12.5).

12.3.5 Securing of Containers

The securing of containers, though often carried out by shore labourers, remains the sole responsibility of the vessel master and their crew. It is important that all rod and wire lashings are sufficiently tightened but not so tight

Figure 12.5 Recommended method for securing unusually shaped cargo on a RORO ship.

Source: Author's own.

as to strain the fittings, containers, etc. It should be clearly ascertained and understood which way the twist handles are put for the locking position. Containers stacked one above the other without the benefit of cell guides must be secured one to the other with twist locks and/or a combination of locating cones, bridging pieces, lashing rods, wires, and shores to prevent any form of shifting. The use of wire, rod, or chain lashings could also be considered. It is also important that only the corner castings are used to secure the container. Deck securing points must be used to advantage. These must never be overloaded. The equipment used must be able to counteract the effect of container jumping, sliding, toppling, and racking. Rigid rods and/or wires are used to counteract jumping, racking, and toppling. Bridge fittings counteract jumping and sliding (Figures 12.6 and 12.7). The following recommendations should be observed when securing containers onboard:

(1) All containers should be effectively secured in such a way as to protect them from sliding and tipping. Hatch covers carrying containers should be adequately secured to the ship.
(2) Containers should be secured using one of the three methods recommended or methods equivalent thereto.
(3) Lashings should preferably consist of wire ropes or chains or material with equivalent strength and elongation characteristics.
(4) Timber shoring should not exceed two metres in length.
(5) Wire clips should be adequately greased and tightened so that the dead end of the wire is visibly compressed.
(6) Lashings should be kept, when possible, under equal tension.

Figure 12.6 Example of container lashing.

Source: Author's own.

Figure 12.7 Example of container lashing mechanisms.

Source: Author's own.

12.3.6 Securing of Heavy Loads

Heavy loads which would normally be stowed fore and aft must be secured against sliding and tipping. The lashing angles against tipping and sliding must not exceed the optimum as stated in the Code. This is shown at Figure 12.8. Heavy items projecting over the ship's side must be additionally secured by lashings acting in longitudinal and vertical directions. If a heavy item does not have suitable securing point, for example, cylinder, then a lashing passing around the item is the preferred option.

Heavy items on deck, which are prone to action from heavy seas, must have additional lashings designed to withstand such impacts. Moreover, the effect of water lifting the cargo must not be ignored or underestimated.

In Australia, Marine Order Part 42 deals with cargo stowage and securing. It aims at timber deck cargo, any large item that will project over the ship's side, items heavier than 100 tonnes, and containers stacked more than one high where the vessel is not designed to do so. Other topics covered by the Marine Order include regulations that are required to be followed when intending to load, crew responsibilities regarding the stowage and securing of cargo, the development and maintenance of the Cargo Securing Manual, safety on deck, and navigational bridge visibility.

Figure 12.8 Securing heavy items.

Source: Author's own.

12.4 DUNNAGE

We have mentioned the word dunnage on a few occasions, without speci-fying what it is and what it is used for. There are generally two types of dunnage which is used onboard ships: permanent dunnage and temporary dunnage. Permanent dunnage is the type that comes with the ship. An exam-ple is steel-corrugated dunnage that is used in reefer compartments. Spar ceiling is another type of permanent dunnage attached to the ship's sides by means of bracket cleats. They protect the cargo from coming in contact with the ship's side where condensation can occur. The gap between the spar ceiling and the hull also allows for air circulation. Reefer containers usually have extruded aluminium 'T' section floors built in. This allows for better air circulation and is a permanent fixture of the container. Temporary dun-nage is that which needs replacing regularly. It comprises hardboard, paper, sawdust, dunnage bags, wood or polystyrene laths, boards, or coir. The dif-ferent types suit different cargoes and require changing once wet, stained, punctured, or oily. Temporary dunnage is very often made of wood and care should be taken to ensure that there are no splinters in the dunnage. In no case should a wet, oily, stained, or smelly dunnage be used. Even new timber has the disadvantage of carrying a strong smell with the risk of staining the cargo in the compartment. Dunnage laid on the tank top normally consists of a double layer, the bottom layer running athwartships to allow drainage to the bilges.

Laying dunnage is not a haphazard function. Dunnage has the purpose of assisting in the solidity of stow and preventing undue damage to cargo in the proximity. In terms of materials which can be used for dunnage, it must be such as to provide protection to and from any or all the following factors: crushing of cargo, dampness, contact with ironwork, and lack of ventilation systems (Figure 12.9). Dunnage may serve the following purposes according to the cargo carried:

- To protect it from contact with water from the bilges, leakage from other cargo, from the ship's side, or from the double bottom tanks.
- To protect it from moisture or sweat which condenses on the ship's sides, frames, bulkheads, etc., and runs down into the bilges.
- To protect it from contact with condensed moisture, which is collected and retained on side stringers, bulkhead brackets, etc.
- To provide air courses for the heated moisture-laden air to travel to the sides and bulkheads along which it ascends towards the uptakes, etc.
- To prevent chafe as well as to chock off and secure cargo by filling in broken stowage, that is, spaces which cannot be filled with cargo.
- To evenly spread out the compression load of deep stowages.
- To provide working levels and protection for the cargo on which labour can operate and serve as a form of separation. To provide access for cooled air round or through the cargo for temperature-controlled requirements.
- To create better friction between the cargo and the ship's floors, thus reducing the risks of moving.

Figure 12.9 Example of temporary dunnage.

Source: Author's own.

Shipowners, and therefore the ships' crew, are legally bound to look after any cargo/passengers in accordance with the *Hague-Visby Rules*. There is an agreement between the shipper/charterer and the shipowner/carrier to that effect. The ship officer has a duty to properly load, handle, stow, carry, and discharge cargo in a similar condition to which it was loaded. As the condition of the cargo when it was loaded is stipulated on the Bill of Lading, any discrepancy at discharging would give rise to disputes. There is no doubt that should there be a discrepancy, then the vessel would be at fault for not exercising its duty of care, thus resulting in lengthy and costly legal procedures. Quarantine requirements sometimes call for a log to be kept when carrying some special cargoes. Some shippers or consignees may ask to see any kept records pertaining to cargo care. These records could be used by the carrier in the event of any dispute as to the way the cargo was looked after. Cargo claims are easily reduced when officers and crew engage in proper caring of the cargo that they are carrying. Above all, any damage to any cargo could mean the loss of the vessel, thus endangering the lives of those onboard. The carriage of coal, for example, requires surface ventilation. Non-compliance with this requirement could lead to a coal fire in the compartment. Cargo that is delivered to the ship in a dangerous condition, that may damage the ship, or other cargo must be rejected and not loaded. The shipper needs to be notified as more suitable cargo may be substituted. Accurate records of observations of cargo condition need to be kept avoiding future claims. So, it is imperative that damage to cargo be prevented at all costs. This will reduce any potential cargo claims and reduce the risks of endangering the vessel and its crew. And this is easily achievable with a little attention from everyone involved.

12.5 CARGO DAMAGE: CAUSES, EFFECTS, AND PREVENTION

Every year, one of the biggest unforeseen and preventable claims that a shipowner face is related to cargo damage. If that trend is consistent over the years, the shipowner will see its P&I contributions being raised together with building up a bad reputation. These negative effects are avoidable. For this to happen, knowledge of the causes of cargo damage and how to prevent such occurrences are of foremost importance. There are various ways in which cargo can be damaged:

(1) *Through temperature change:* Some cargoes are carried at ambient temperature. However, a too hot temperature will cause it to melt (chocolate), to disintegrate (through chemical reaction), or to crack and break. Other cargoes if in a very low temperature might freeze or crack. Temperature control through ventilation is a good way

of preventing this. Attention should be paid to the stowing. Do not stow in places where temperature extremes are possible.

(2) *Through chafing and rubbing:* Due to the vessel's movement or vibration, cargoes can rub against each other or the ship structure. This will result in damage to the packaging and, eventually, the cargo itself. It can be avoided by a compact stow or putting chocks between the packages or separation.

(3) *Through contamination:* Cargo can get damaged in various ways by contamination – one of which is by dust. Dusty cargo must be loaded first. Cargo in the same compartment must be covered while dusty cargo is being loaded. Special attention must be paid to stains on tank tops as these could also contaminate cargoes. The use of dunnage could reduce that risk.

(4) *Through cargo mixing:* This is another type of cargo contamination, which occurs when two cargoes are stowed together and the separation between them is inadequate. Large plastic sheets should be used to separate them. In other instances, the contamination may not be physical, but olfactive, like tainting of fruits.

(5) *Through rust:* Moisture deposited on steel will turn the cargo rusty, with risks of it being rejected at the discharging port. This moisture can be from different sources: internal (other cargo leaking, sweat) or external (compartment not watertight). Ventilation could be the solution, though sometimes it could be better not to ventilate. Proper planning will help reduce this risk.

(6) *Water damage:* Could be the biggest threat to cargoes. Water can cause rust, mould, and discolouration of cargo, caking, or sweat. Water-tight integrity of the compartment must be always maintained. Proper ventilation will prevent sweat. Proper planning is required. Some class IV cargoes, for example, are dangerous when wet.

(7) *Improper lashing:* Inadequate lashing can not only mean damage to cargo, or its loss, but also the loss of the vessel and its crew. Always lash for the worst possible condition and check regularly.

(8) *Pilferage:* Though this activity has been made more difficult with the advent of containers, it is still a major concern. Valuable goods are to be loaded last, preferably in lockers in the presence of watchmen. Do not advertise the goods and keep an accurate tally. Watch out for the condition of the packages, as they are loaded. If suspicious, do not load. Have adequate lighting if loading at night. Remember that containers of valuable goods have disappeared too.

(9) *Through mechanical mishandling:* Using the wrong tool to handle the cargo can at times be detrimental. In trying to speed up operations, this could happen and a claim for cargo damage will follow. Have the right gear ready for use, and in good working condition.

(10) *By rats, mice, and vermin:* Rats and mice destroy some cargo to the extent that they are rejected at the discharging port. Once the vessel is fumigated, every effort should be made to keep the rats away. If they are loaded with the cargo (with wood or pallets, for example), the hold should be cleaned and fumigated after discharge. However, fumigation itself or its residues might pose a hazard to some cargo.

12.6 TEMPERATURE-CONTROLLED REEFER CARGO

Reefer cargoes can be split into three main categories:

(1) Frozen cargo.
(2) Chilled cargo.
(3) Temperature-regulated cargo.

Frozen cargo is carried in the hard frozen condition, which means that a temperature of at least -20°C (-4°F) must be attainable. Most frozen commodities are carried at a temperature below -7°C (19.4°F) when no microorganism growth is possible. For example, frozen lamb carcasses. It can be carried either in reefer containers or in cold chambers. Chilled cargoes are commodities where the outside has been frozen hard, but the inside remains unfrozen. Chilled meat is tastier – thus more expensive – than frozen meat. The usual temperature range is between -2°C (35.6°F) and -3°C (37.4°F). This small range calls for great care from the ship's personnel. In chilled cargoes, the growth of microorganisms is only slowed down. Thus, chilled cargo cannot be kept in that condition for a long period of time. Its carriage must be made within time limits. If CO_2 is injected into the compartment, then the duration is prolonged. For example, chilled beef or lamb. It is usually carried in containers, though its transportation in cool chambers has also proved successful. Temperature-regulated cargoes are those which are carried at a temperature which restricts processes such as ripening. Commodities require different carrying temperatures. Goods such as apples can be carried as low as 1°C (33.8°F), whilst citrus fruits are carried at 10°C (50°F). Normally in the carriage of fruits, the latter will give off CO_2 which, if in low concentration, is beneficial as it slows down the ripening process. The refrigerated cargoes mentioned earlier can be carried in two ways:

(1) Onboard custom-built reefer ships or general cargo ships with limited reefer spaces.
(2) In containers onboard ships capable of accommodating reefer containers.

One very important feature of a reefer ship is its speed. Once the cargo is loaded, it has to be despatched to destination rather urgently. The other important feature of a reefer ship is its refrigeration plant and cold chambers. Reefer ships normally have three or more decks which are subdivided so that cross-contamination is not possible and that a range of cargo with different carrying temperature can be loaded. Table 12.1 shows a segregation table involving commodities that are carried at regulated temperatures. The segregation is contamination related. This means that despite some of the fruits having the same carrying temperatures, they have to be segregated because of the possibility of tainting. The cargo temperature is usually controlled by a ventilation system forcing cold air in. Air is made to pass over an evaporator and circulated in the cargo compartment. The refrigerated chambers normally have temperature-monitoring devices linked to the control room. The components of a refrigerant plant are illustrated in Figure 12.10. An understanding of the principles of operation of the plant is part of another module.

Care should be taken so as not to damage the insulation, which is around the compartments. As this insulation is sometimes highly flammable, under no circumstances should any hot work (especially welding) be carried out without prior checking the surroundings. Any escape of cold air will cause undue stress on the compressors and evaporators. Sealing of the compartment is common practice. Also, during cargo operations, should there be a break in the loading/discharging, the compartment should be closed so as to avoid any loss of cool air. The refrigeration plant should be stopped, also.

12.6.1 Types of Reefer Containers

There are two types of reefer containers: independent and ship-dependent containers. The ship-dependent type is today extremely scarce. Independent reefer containers have their own refrigeration plant but require electrical power from the vessel. Special power points and, usually, a dedicated generator will provide power for these containers. The temperature is regulated before packing the container and is automatically controlled. In modern reefer containers, there is even a spare compressor which is carried in case the main one breaks down. There is hardly any need for maintenance from the ship's crew. A graphical record is always kept as part of the container equipment. Though the monitoring of the temperature is done automatically, it is good practice to carry out checks daily. The location of power points, action of sea and waves, accessibility to the container and above deck stowage are points that should be considered when planning the stowage of reefer containers. It is important to connect power soon after loading to prevent container warming up. Always check contents – sometimes general is shipped in spare reefer containers. When carrying reefer containers in cold chambers, it is of great importance that the cold air is allowed to circulate around and, in some cases, through the cargo. Deep frozen cargo can be stowed in a solid block as it is only necessary to keep the boundary cold to ensure the

Table 12.1 Commodities Segregation Table for Refrigerated Cargoes

	Apples	Bacon	Beef (chilled)	Beef (frozen)	Butter	Cheese	Fish (frozen)	Grapes	Mutton	Oranges	Pork	Peaches	Plums	Potatoes	Vegetables	Lobster
Apples	X	N	BR	BR	N	N	SR	Y	N	Y	N	Y	Y	Y	Y	N
Bacon	N	X	SR	Y	SR	Y	SR	SR	SR	N	Y	SR	SR	SR	Y	SR
Beef (chilled)	BR	SR	X	Y	Y	SR	Y	Y	Y	N	Y	Y	Y	SR	Y	SR
Beef (frozen)	BR	Y	Y	X	Y	SR	Y	Y	Y	N	Y	Y	Y	SR	Y	SR
Butter	N	SR	Y	Y	X	SR	Y	Y	Y	N	SR	Y	Y	N	Y	BR
Cheese	N	Y	SR	SR	SR	X	SR	Y	SR	N	SR	SR	SR	SR	SR	N
Eggs	N	Y	Y	Y	Y	N	X	Y	Y	N	Y	SR	SR	N	Y	SR
Fish (frozen)	SR	Y	Y	Y	Y	N	SR	Y	Y	N	Y	SR	SR	SR	SR	Y
Grapes	Y	Y	Y	Y	Y	SR	Y	X	Y	Y	Y	Y	Y	Y	Y	Y
Mutton	N	Y	Y	Y	Y	SR	Y	Y	X	N	Y	Y	Y	SR	Y	SR
Oranges	Y	N	N	N	N	N	N	Y	N	X	N	N	Y	Y	Y	N
Pork	NY	Y	Y	Y	SR	SR	Y	Y	Y	N	X	Y	Y	BR	Y	SR

Figure 12.10 Typical refrigerated 'reefer' container.

Source: Author's own.

whole stow is cold. Chilled cargo is most frequently carcass and as such there will be sufficient space between the carcasses to ensure a good circulation of cool air. Temperature regulated cargoes such as fruit and vegetables are often transported in bins or cases. Because of the continued life of these products after they have been picked, it is necessary for a good airflow to ensure they can 'breathe'. This means that boxes should have ventilation holes and there should be enough space to allow air circulation between the boxes. Refrigerated cargoes are normally high-value cargoes. Because of this, any failure by the ship's staff to care for the cargo in a correct manner can be an extremely costly exercise. The maintenance of correct carrying temperatures cannot be overemphasised, as these are the prime responsibility of the ship staff.

12.7 SWEAT: CAUSES, EFFECTS, AND PREVENTION

Sweat is one of the main causes of cargo damage. It can cause cargo to rust, go mouldy, germinate, rot, or simply become unusable. It could also cause damage to the vessel and its structure. Sweat is caused by the condensation of moist saturated air on a cooler surface. Some definitions are useful in understanding sweat.

- *Dew point:* The dew point is defined as the temperature at which air cannot absorb any more water vapour, that is, the air has become fully saturated.
- *Saturation:* This infers that a parcel of air has absorbed its maximum amount of moisture for any given temperature. Raising the temperature of the air will allow for more moisture to be absorbed.
- *Relative humidity:* This is defined as the ratio between the actual moisture content of the air and the maximum moisture content that the parcel of air can contain.
- *Hygroscopic cargo:* Hygroscopic cargoes are those which can absorb or give off moisture. This includes timber products, grains, and ores.
- *Non-hygroscopic cargo:* These are cargoes which do not absorb or give off moisture, that is, the opposite of hygroscopic cargoes.
- *Ship sweat:* This results when water droplets are deposited on parts of the ship structure when air is cooled below its dew point. The condensation occurs because the ship structure is cooler than the dew point of the surrounding air.
- *Cargo sweat:* This results when water droplets are deposited on parts of the cargo when the surrounding air contacts cooler cargo. The temperature of the cargo must be lower than the dew point of the surrounding air.
- *Moisture control:* Moisture control is usually the most serious air control problem aboard a general cargo ship. Moisture can damage cargo in two ways:

(1) By causing condensation
(2) By causing the cargo to germinate or cake in the case of hygroscopic cargoes

Controlling damage to non-hygroscopic cargo can be easily controlled. This is because non-hygroscopic cargo does not absorb, retain, or reject moisture. Thus, any sweating occurring will be in the form of ship sweat dripping onto the cargo. Therefore, a way of controlling the dew point of the surrounding air and the temperature of the ship structure must be found. Dew point is a function of moisture content. Condensation will only occur if the air is saturated, and it is cooled below its dew point. If the moisture content is reduced, then the air is not near its saturation point. And condensation is unlikely to occur. So, it can be said that by reducing the moisture content of a parcel of air, the risks of condensation are reduced. Alternatively, when it comes to hygroscopic cargoes, the moisture content of the surrounding air and that of the cargo have to be taken into account. Both are capable of absorbing, retaining, or giving off moisture. This exchange of moisture is necessary to achieve a balance in moisture content and will result in condensation. When hygroscopic cargoes are carried, several factors are worthy of consideration:

- The moisture content should be kept as low as possible (hygroscopic cargoes should not be exposed to rain before loading); and
- It is important to keep hygroscopic cargo cool because the presence of heat causes the dew point of the surrounding air to rise, which in turn means there is a high risk of condensation.

The solution in preventing damage to hygroscopic cargoes lies in maintaining the dew point of air in the cargo space below the dew point of the air in and immediately around the cargo.

12.7.1 Examples of Ship Sweat and Cargo Sweat

Imagine a cargo hold surrounded by seawater at a temperature of 15°C (59°F); therefore, the steel structure would also be close to that temperature. Now imagine if, inside the hold, some cargo was loaded in a cold country at a temperature of about 10°C (50°F). Now, if air from outside at a temperature of 18°C (64.4°F) and a dew point temperature of 13°C (55.4°F) is allowed in, cargo sweat will occur on the surface of the cargo. However, there will be no ship sweat because the air will not be cooled below its dew point. If the temperature of the ship structure were 12°C (53.6°F), then ship sweat would result (Figure 12.11). From the example, it is apparent that certain information is required before it is possible to decide whether or not to ventilate the cargo space:

- Temperature and dew point of the outside air.
- Temperature of the cargo surface.
- Temperature of the steel structure inside the cargo compartment.
- Temperature and dew point of the inside air.
- The moisture content of hygroscopic cargoes if applicable.

Sometimes ventilation is necessary to achieve a desired outcome. The basic rules for ventilation are as follows:

(1) For hygroscopic cargo:

   (a) Warm to cold: ventilate vigorously initially.
   (b) Cold to warm: ventilation is not necessary.

(2) For non-hygroscopic cargo:

   (a) Cold to warm: no ventilation is needed even if cargo sweat is likely to occur.

In cold to warm conditions, ship sweat is inevitable, but cargo is unaffected unless condensation drips back on to the stow. These are not hard and fast

Figure 12.11 Cargo damaged by seawater.

Source: Author's own.

rules as, at times, there will be condensation whether the space is ventilated or not. Thus, as a guide, the general rule that can apply to ventilation is as follows: ventilate if the dew point outside is *lower* than or equal to the dew point inside the compartment.

12.8 CARGO CONTAMINATION

Contamination is another cause of cargo damage. It can occur in many ways with varying effects. Contamination can frequently occur on tankers and product carriers. Two types of oil which are incompatible could be mixed accidentally, rendering the cargo unusable or even dangerous. This is very important on product carriers, which carry vast amounts of small parcels of cargo. The purity of these parcels must be maintained. This is achieved by dedicating special lines to the types which are highly incompatible. In other cases, the lines are drained and cleaned thoroughly before handling a different type from the one which had been handled previously. The tanks are cleaned scrupulously, and a certificate of cleanliness is issued before any loading operations can proceed.

12.8.1 Contamination by Taint

Tainting is another aspect of contamination. Imagine carrying bags of onions together with textile products. These will be absorbing that onion smell and would thus lose their commercial value. Any cargo which gives off a strong smell is liable to taint absorbent material. This can be prevented by stowing the cargoes in different compartments or sealing one of them in plastic or polystyrene wrap.

12.8.2 Contamination through Fumigation

Cargoes can become contaminated through fumigation. Care should be taken when using fumigants. These should be of the type that will not damage the cargo in the compartment. If fumigating an empty compartment, residues from the fumigant may still be harmful to the cargo to be loaded. After fumigating, it is advisable to ventilate the compartment to some extent.

12.8.3 Contamination by Water

Water is known to be one of the biggest contaminants. The damage resulting from water contamination ranges from germination, swelling, discolouration, or even dissolution. Keeping the compartment watertight/preventing the cargo from contact with ship's sweat are two of the measures that could be taken to prevent water contamination.

12.8.4 Cross-Cargo Contamination

Cargo can be contaminated by other cargoes. This is particularly the case when loading general cargoes, part of which consists of dry powdery solids such as fertilisers and flour or grain and granules.

12.9 CARGO DAMAGE SURVEY PROCEDURE

Part of the job of the cargo officer when engaged in cargo work is to make a note of the condition of the cargo being handled. This is even more important if the cargo is damaged or seems to be in an unusual state. Any abnormal condition of the cargo should be noted in the cargo logbook. If during cargo operations cargo is accidentally damaged, the duty officer must make a note of it – noting the time such incident happened, the amount of the cargo damaged, the extent of the damage and any marks on the cargo. The chief officer's attention should be drawn immediately, and a cargo damage report must be made. Witnesses to the incident could be called in to give their version of the incident. The stevedores involved in the cargo handling must also be part of the report. Depending upon the extent of the damage, independent surveyors and/or the corresponding P&I surveyors must be called in. Steps must be taken to prevent further damage to the cargo. This information must be filled in the cargo logbook. The independent marine surveyor is the expert in making any cargo damage reports (Figures 12.12 and 12.13). In their report they will include the following items:

- Date, location of requested survey
- Details of persons requesting the survey

- Details of carrying vessel.
- Description of the shipment.
- Facts regarding the stowage.
- Details of loading/discharging.
- Names and addresses of shipper and consignee.
- Extracts from ship's logbook, cargo logbook, note of protest, or any other relevant document.
- Full description of the damage.
- Details of any action taken to reduce damage.

Figure 12.12 Independent cargo inspector.

Source: Author's own.

Figure 12.13 Independent cargo inspector report.

Source: Author's own.

The surveyor has the right to expect the full cooperation of the vessel's crew and to be provided with information that would be the base for their report. Thus, a detailed account of the events surrounding the damage must always be kept for reference.

12.10 SUMMARY

No matter what, the cargo officer must take all actions to limit the damage to cargo. This might even include stopping cargo operations. The damaged cargo could be repacked and discharged in an unconventional way. It depends upon the circumstances and the extent of the damage. No damage to cargo should go unnoticed. If damaged goods are loaded and not brought to the attention of the master, they might find themselves facing major cargo claims, as there will be no evidence that the damage occurred elsewhere than on the vessel.

Chapter 13

Safe Handling, Stowage, and Carriage of IMDG Cargo

13.1 INTRODUCTION

Every cargo is special to its shipper or carrier; however, there is no cargo which is more special than dangerous cargo! But what exactly is a dangerous cargo? Why is it so important and what are the regulations set in place to facilitate its handling, stowage, and carriage? The term 'dangerous cargo' means any substance which, because of its inherent properties, can be harmful to any person, animal, or the environment. Some materials may also be affected if they are exposed to different types of dangerous goods. What makes dangerous goods special are the risks that are involved in their handling, stowing, and carrying. The risks involved are very high, numerous, and varied. For this reason, strict rules have been set up and ship personnel must abide by them. Rules governing the packing, handling, stowage, and carrying of dangerous goods are laid down by IMO in the publication *International Maritime Dangerous Goods Code*, more commonly known as the IMDG Code. The classification, packaging, and stowage of dangerous goods must be in accordance with any legislation which may be enforced in (a) the country of origin; (b) the country of destination; (c) any country which it has entered; and (d) the flag state of the carrier. Regulation pertaining to the carriage of dangerous goods applies to all ships but does not apply to the ships' own stores and equipment.

To be conversant with the carriage of dangerous goods, seafarers must understand fully some of the most common terms used in the field. These are addressed throughout this chapter. The current IMDG Code for calendar year 2022 is either the 2018 Edition, incorporating amendments 39–18, or the 2020 Edition, incorporating amendments 40–20. Both the 2018 Edition and the 2020 Edition can be used in 2021. It is now in two A4 paperback volumes, replacing the old four-volume loose-leaf editions. The new reformatted code came into force on 1 January 2001, with a 12-month implementation phase. The IMDG Code supplement has also been produced and includes a revised Medical First Aid Guide and the new mandatory *International Code for the Safe Carriage of Packaged Irradiated Nuclear Fuel, Plutonium and High-Level Radioactive Waste Onboard Ships*, 2001

DOI: 10.1201/9781003354338-15

(the INF Code) as amended by resolutions MSC.118(74), MSC.135(76), MSC.178(79), and MSC.241(83). It is a legal requirement that the shipper must provide the vessel with all relevant information pertaining to the dangerous goods being loaded. However, discrepancies may creep in. Usually, both the technical name and the UN Number are provided. This allows for cross-checking. If the UN number is known, then go straight to the dangerous goods list in Volume 2 – the UN numbers are in order. In our example, the UN number is 1360. If only the name is known – calcium phosphide – refer to the index in Volume 2 to obtain the UN number. If the name and number are known, this index can also be used to cross-check them. In the dangerous goods list now, working across the columns will give the information for that substance, its class, subsidiary risks, packing group, and instructions – all of which are found in the code by following the sections indicated. Stowage and segregation are given, as are the properties and observations of the substance. Most importantly, the EMS table is given; in this case, 4.3–02. Going to these sections, you will see the appropriate emergency schedule and reference to consult the first aid guide.

13.2 STRUCTURE OF THE IMDG CODE

Volume I (Parts 1, 2 and 4–7 of the Code) includes the general provisions; definitions, training, and the classification; packing and tank provisions; consignment procedures; the construction and testing of packaging, intermediate bulk containers, large packaging, portable tanks, and road tank vehicles; and transport operations. Volume II contains Part 3 of the Code which incorporates the Dangerous Goods List (presented in tabular form); the limited quantities exceptions; the index; and appendices. The Supplement contains the Emergency Procedures; Medical First Aid Guide; Reporting Procedures; Packing Cargo Transport Units; Safe Use of Pesticides; INF Mandatory Code; and Appendix. The dangerous goods listed in the IMDG Code do not all have the same danger risks. Some might be explosives, while others can be poisonous. Others might be dangerous because they are radioactive. For this reason, dangerous goods are divided into classes. There are nine classes, some of which are further divided into subclasses:

13.2.1 Subclass

This means that some substances in a particular group can be put into a smaller category. For example, some poisonous substances can be subclassed as being infectious substances.

13.2.2 The UN Number

The simplest way of identifying a cargo, which sometimes can have a very complicated chemical name, is by a four-digit number. Each substance listed in the IMDG Code is assigned a specific number. This is known as the UN Number of that substance. It reduces the risks of making a mistake when spelling out the technical name of that cargo. Volume II of the code lists all the dangerous cargoes in order of their UN number.

13.2.3 General Index

In the case where the technical name is known, and there is doubt about the UN Number, the General Index, which is part of the IMDG Code Volume II, is consulted and the UN number is extracted for further information about a particular cargo.

13.2.4 Segregation

Some cargoes, because of their chemical and physical properties, have a tendency to react with other substances or parts of the vessel. Thus, they have to be stowed far away from each other. The term 'segregation' is used to denote this practice. There are various grades of segregation. This will be dealt with later in this chapter.

13.2.5 Labelling

Grouping dangerous goods in classes is good. But once the goods are packed, one would not know what the package contains unless some big, colourful, and explicit labels are affixed to it. For each class of DG, there is at least one main label. The diamond-shaped label has a drawing on its top half and the class number at its bottom corner. In some cases, the type of DG is also written on the label.

13.2.6 Subsidiary Risk Label

Some cargoes have more than one risk. An example would be an inflammable substance that could also be corrosive. This means that the cargo has a subsidiary risk. Thus, precautions must be taken against two types of dangers. The way the cargo is labelled in this case will be that it will carry an additional label to that of the main risk. A class label without its class number at its bottom corner can identify the subsidiary risk label.

13.2.7 Marine Pollutant

Some dangerous goods, apart from having main and subsidiary risks are also marine pollutants. This means that, if they are spilled, they will pollute the marine environment. Again, there are two levels of pollutants: a marine pollutant and a severe marine pollutant. These cargoes carry a Marine Pollutant Mark. The triangular shape makes the mark different from a class label.

13.2.8 Packaging Group

Some dangerous goods have, for packing purposes, been divided into three categories (packaging groups) according to the degree of danger they present: great danger (packaging group I), medium danger (packaging group II), and minor danger (packaging group III).

13.2.9 Emergency Schedule Number

The Emergency Schedule Number (EMS) is contained in the Supplement to the Code. The table should be consulted in the event of an emergency involving a particular dangerous good.

13.2.10 Medical First Aid for Use in Accidents Involving Dangerous Goods Table Number

This contains advice on Medical First Aid for use in Accidents Involving Dangerous Goods. The Medical First Aid Guide Tables are found in the Supplement to the Code. Refer to the IMDG Code for more details.

13.3 CLASSIFICATIONS OF DANGEROUS GOODS

There are nine classes of dangerous goods, with some of them being further divided into subclasses. They are classed according to their chemical properties, reaction, and risks associated therewith. The marking of these classes must be done according to the Code. This reduces the risk of confusion and misunderstanding while providing a universally agreed standard.

13.3.1 Class 1: Explosives

This covers diverse hazards ranging from safety class ammunition to those which have a mass explosion risk. This class is usually subject to stringent legislation and port bye rules. Explosives can only be carried in conjunction with the regulations stipulated in the Code:

(1) When explosives are to be handled, unnecessary persons should not be allowed nearby.
(2) The cargo should be loaded last and discharged first.
(3) Detonators should be stowed away from explosives. All necessary measures should be taken to prevent any movement of the cargo during the voyage. Ventilation may be necessary, while all electrical circuits around the cargo must be isolated.
(4) Firefighting appliances should be permanently rigged. If possible, the cargo should be stowed as far away from the accommodation

as possible. Explosives (except ammunition) which present a serious risk must be stowed in a magazine, which must be securely closed when at sea. Such explosives must be segregated from detonators.

(5) Electrical apparatus and cables in any compartment in which explosives are carried must be designed and used so as to minimise the risk of fire or explosion.

13.3.2 Class 2: Flammable Gases

According to their properties or psychological effects, which may vary widely, gases may be explosive, inflammable, poisonous, corrosive, or oxidising substances or may possess two or more of these properties simultaneously. Some gases are chemically and psychologically inert. Such gases as well as other gases may be regarded as non-toxic but may be suffocating in high concentrations. Some gases have narcotic effects or may evolve into poisonous gases when involved in a fire when pressure build-up may cause them to explode. Some substances are liable to alter under transport conditions, to combine or react themselves to cause dangerous liberation of heat or gas, resulting in pressure on the receptacle. These substances should not be transported unless they are properly inhibited or stabilised.

13.3.3 Class 3: Flammable Liquids

The danger associated with inflammable liquids is the escape of inflammable vapours (some of which could be toxic) prone to substances having a low flashpoint which are naturally volatile. The vapour could mix with air leading to an explosion or catch fire through becoming ignited by a spark or flame. *Precautions:* These substances should be stowed away from naked lights, fires, or any source of heat and packaging should serve to protect the contents against external source of ignition. The flammable substances could be miscible or immiscible with water, a point to note when firefighting.

13.3.4 Class 4.1: Flammable Solids

Solids that readily ignite: some may explode unless kept in a saturated condition with water or some other liquid might make the substance become dangerous. Keep away from any source of ignition.

13.3.5 Class 4.2: Flammable Material Liable to Spontaneous Combustion

They can either be solids or liquids. Such substances should be carefully watched for any rise in temperature. Those which ignite immediately in contact with air are especially dangerous. Vegetable fibre should be kept free from contamination by oil or water. This should not be loaded as self-heating may commence some days or weeks later.

13.3.6 Class 4.3: Flammable Materials that Are Dangerous When Wet

These are substances which, when in contact with water, emit flammable gases. All substances in this class must be kept dry. In some cases, the gases may be toxic. Some of these are liable to spontaneous ignition due to heat liberation by the reaction. The characteristics of each substance in this class should be closely studied and no cargo likely to interact packed in the same compartment or container. Refer to Appendix 9.

13.3.7 Class 5.1: Oxidising Substances

These substances are not combustible on their own but possess the ability of making combustible material burn easily. Upon burning oxygen is given off, thus increasing the intensity of the fire. Mixing some oxidising substances with combustible material can create a highly flammable mixture, capable of being ignited by friction alone. Toxic gases could be released if oxidising substances are allowed to react with some acids.

13.3.8 Class 5.2: Organic Peroxides

Organic peroxides are both oxidising agents and inflammable and will burn readily, sometimes with explosive force. All may decompose with heat and, in general, are the most unstable of substances. Some evolve oxygen naturally and are packed in receptacles which are provided with a means of ventilation. Carriage of such substances may require temperature control.

13.3.9 Class 6: Poisonous Substances

Poisonous substances are liable to cause death or serious injury to human health if swallowed, inhaled, or by skin contact. Nearly all toxic substances give off toxic gases when involved in a fire. Breathing apparatus sets and protective clothing should be readily available in case of damage to packages.

13.3.10 Class 7: Radioactive Substances

The care and handling of radioactive substances varies widely. Very stringent precautions are taken to ensure the safe packaging of radioactive substances, and these are all within internationally agreed standards. Careful study of all ports regulations and documentation of goods in this class is of the highest importance. Staff should seek guidance through consulting the appropriate statutory regulations, or the authorities concerned, whenever necessary.

13.3.11 Class 8: Corrosives

Some solids or liquids, namely corrosives, when in their original state are capable of damage living tissue. The escape of such substances is sufficiently volatile to evolve vapour irritating to the nose and eyes. A few substances may produce toxic gases when decomposed by high temperatures. In addition, poisoning may result if corrosives are swallowed or if vapour is inhaled; some of them may even penetrate the skin. All have a more or less destructive effect on materials such as metals and textiles.

13.3.12 Class 9: Miscellaneous

Miscellaneous dangerous substances are a class which contains substances which, although dangerous, have not been allocated to any other class. It includes substances which cannot be brought under any of the more precisely defined classes because they offer a particular danger which cannot be properly covered by the regulations for the other classes, or which present a relatively low transportation hazard. It should not be automatically assumed that substances in this class are 'less hazardous'.

13.4 MARKINGS OF DANGEROUS GOODS

13.4.1 Class 1: Explosives

The diamond-shape label is of orange colour and shows the class number at its bottom corner. Apart from that, the appropriate division number and compatibility group numbers may also be shown in the centre of the label (Table 13.1).

Table 13.1 Class 1: Explosives

13.4.2 Class 2: Gases

The label can be white, green, red, or yellow depending upon its subclass. The various symbols and legend remove any doubt about the type of dangerous good being handled (Table 13.2).

Table 13.2 Class 2: Gases

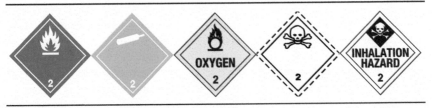

13.4.3 Class 3: Flammable Liquids

The entirely red label together with its legend makes it easy for identification (Table 13.3).

Table 13.3 Class 3: Flammable Liquids

13.4.4 Class 4: Flammable Solids

Because of the various risks that exist with flammable solids, sub-classification is necessary. The labels are of different colour with the same symbol but different legends. As usual, the class number is found at the bottom of the label (Table 13.4).

Table 13.4 Class 4: Flammable Solids

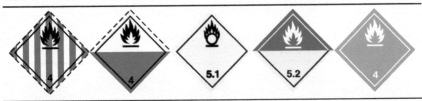

13.4.5 Class 5: Oxidising Substances and Organic Peroxides

These two subclasses have a different label, different legend, and different subclass number (at the bottom corner), but the same symbol (Table 13.5).

Table 13.5 Class 5: Oxidising Substances and Organic Peroxides

13.4.6 Class 6: Poisonous Substances and Infectious Substances

The white label carries different symbols, different legends, and the same class number at the bottom corner (Table 13.6).

Table 13.6 Class 6: Poisonous Substances and Infectious Substances

13.4.7 Class 7: Radioactive Substances

Labels can be either all white or white and yellow. The class number is the same for all three types of labels which characterise different radioactive categories (Table 13.7).

Table 13.7 Class 7: Radioactive Substances

13.4.8 Class 8: Corrosives

There is no subclass for corrosives. The black and white label and its symbol cannot be mistaken for any other label (Table 13.8).

Table 13.8 Class 8: Corrosives

13.4.9 Class 9: Miscellaneous Substances

This also is a black and white label, but without any symbol. It has the class number at the bottom corner. Apart from the class label, the subsidiary risk label and the marine pollutant mark, any dangerous substance can also show an elevated temperature mark. If fumigation is being carried out, then a fumigation warning sign must be displayed. Substances with multiple hazards must conform to a table of precedence contained in the Code. The precedence of hazard table indicates which of the hazards should be regarded as the primary hazard. The carriage, stowage, and discharge of dangerous goods involve several risks associated with each different class of cargo (Table 13.9). As a general rule, the following recommendations should always apply before attending to the specific needs of each class:

(1) Dangerous goods should be stowed safely and appropriately according to the nature of the goods.
(2) Incompatible goods must be segregated from one another.
(3) Goods which give off dangerous vapours shall be stowed in a well-ventilated space or on deck.
(4) Where flammable liquids or gases are carried, special precautions are to be taken where necessary against fire or explosion.
(5) Substances which are liable to spontaneous combustion or heating shall not be carried, unless adequate precautions have been taken to prevent the outbreak of fire.

Table 13.9 Class 9: Miscellaneous Substances

13.5 SAFETY PRECAUTIONS AND PROCEDURES DURING CARGO OPERATIONS

As stated above, the carriage of dangerous goods can only be done by following the rules laid down in the IMDG Code. Furthermore, Marine Order Part 41 describes the procedures to be followed if dangerous goods are to be loaded on a ship in an Australian port or onboard an Australian vessel:

(1) Adequate warning identifying the dangerous cargoes must be given to persons handling the cargoes and to crew members.

(2) The vessel must be fitted with an emergency alarm capable of emitting an audible alarm.

(3) There must be in place a notice prohibiting smoking.

(4) There must be in place a fire hose in a state of readiness for immediate use.

(5) If water is unsuitable, portable fire extinguishers must be available.

(6) Smoking, hot work or flame work is not allowed in the vicinity where dangerous goods are being handled.

(7) Lighting in place must be electric and so constructed that it will not short-circuit or spark.

(8) Bunkering is not allowed when Class 1 goods are being handled.

(9) Class 1 goods must always be protected from direct sunlight or from becoming wet.

(10) When handling Class 1 goods, VHF or UHF transmitters can only be used if more than 2 metres (6.5 ft) from the cargo.

(11) Radio transmitters must not be used except if they are more than 100 metres (328 ft) from class 1 cargoes when they are being handled.

13.6 ACTIONS IN THE EVENT OF LEAKAGE OR SPILLAGE

In the event of any leakage or spillage, the prescribed must be advised of the incident immediately. Any dangerous goods which have escaped must not be touched and other cargo in the same space must not be handled. No pumping of bilges is allowed unless risks of pollution are deemed to be non-existent. The space where the spillage has occurred must be cordoned off until the prescribed person has authorised resumption of normal duties. However, a ship's officer, an approved chemist, or a surveyor is allowed in that space to attend to the emergency. More importantly, the EMS and MFAG must be consulted at once and recommendations followed.

13.7 PROCEDURES TO FOLLOW WHEN SHIPPING DANGEROUS GOODS

When preparing to load dangerous cargoes for carriage by sea, the shipper must state their intention to ship dangerous goods in advance by completing the various forms required in accordance with Marine Order Part 41 (Marine Order 41/1 and Marine Order 41/2) for vessels under Australian Flag State jurisdiction

(Figures 13.1–13.7). The vessel must be advised that the booking list contains dangerous goods. The chief officer is then required to prepare a pre-stowage plan showing the proposed location of the dangerous goods. Factors that are taken into consideration when doing so include the class, amount, the risks involved, and packaging and stowage requirements for each component of the dangerous goods manifest. The vessel must then advise the local Port Authorities of its intention to load dangerous goods. This is done using form Marine Order 41/3. Once the form has been received and processed by the Port Authority, a prescribed person will visit the vessel and approve the loading plan. Once the loading plan is fully approved, and authorisation is granted, the cargo may be loaded under strict surveillance. A final stowage plan is then made. The plan must show the exact location of the dangerous goods, the type of dangerous goods, and the amount of dangerous goods loaded. A dangerous goods manifest is prepared for the vessel to hand in to the various authorities at the next port of call.

ANZDL ☑☑
Australia-New Zealand Direct Line

Level 9, 234 Sussex St
Sydney, NSW, 2000
Fax: 61 2 9261 1993
Tel: 61 2 9267 1911

Page 1 of 1

FACSIMILE TRANSMISSION

TO: DEFLOG- LLOYD MCCALLUM   FAX NO: 99555795

CC: John Shears/ Patrick Darling Harbour
Stephen Lacey;
FROM: Manjur Khan

DATE: 8th August 2000

SUBJECT: DARLING HARBOUR - LOADING OF CLASS 1.3C

Dear Lloyd,

This is to advise you that the **Direct KEA, V.480** will be berthing at no.4 Darling Harbour, for discharge and loading operations.

The Vessel's current ETA Sydney is 23/08/00, and **ETD 24.08.2000**

Mr. John Shears, Operations Superintendent, at Patrick Darling Harbour has assured me that there shall be a clearance of 180 mts from the Direct Kea and this will facilitate the loading of the container containing Haz class 1.3C.

Following are the guidelines given to me by Mr. Davis at SPC:
7.5 MT NEQ = 125 MTRS separation
12.5 MT NEQ = 148 MTRS separation
20 MT NEQ = 171 MTRS separation

Please do not hesitate to contact me if you have any further queries.

Thanking you,

Best regards

Manjur Khan

ANZDL Management Services Austracia Pty. Ltd
A.C.N. 001 000 506

Figure 13.1 Example notification of intent to load dangerous goods.

Source: Author's own.

DEFLOG International Pty Limited   ABN 15 071 114 139

Level 2, 161 Walker Street, North Sydney NSW 2060 Australia

Tel: +61 (0)2 9955 5822   **Fax:** +61 (0)2 9955 5795   **Email:** info@deflog.com.au

Fax Transmission

TO:	AMSA Sydney	ATTN:	James McLeod	0292820750
TO:	WorkCover Authority	ATTN:	Philip Owens	0293706105
TO:	ANZDL – Sydney	ATTN:	Manjur Khan	1800818731
TO:		ATTN:	Agnes Jakovlej	1800818731
TO:	Sydney ports corporation	ATTN:	Robert Powell	0292964812
DATE:	18 august 2000	OUR REF:	20564	

From: Stephen R. Lacy   No of pages incl. this one: 2

Subject: *Export of 1 x Class 1.3C (Red Line) container ex Sydney – No.4 North Darling Harbour*

It is our *intention to export* up to 9,285 kgs NEQ of PSN; Powder, smokeless IMO Class **1.3C UN0161** on behalf of ADI Limited, Operations Group, Mulwala NNSW from Patricks Terminal - No.4 North Darling Harbour.

Vessel: **"Direct Kea" V.48ON**   **ETD: 24 August 2000**
Nbr of Cartons: 1,238 cartons.

Attached is a copy of the M041/1 &2 form signed by the shipper.

Container No. **CRXU 2111911** has been surveyed by the AMSA Sydney office and certified to be suitable for the carriage of Class 1 cargo in accordance with the requirements specified in the IMDG Code Class 1 Explosives Section 12.3.

The empty container will be delivered to the ADI Mulwala NSW Facility for loading on Thursday 24 August 00 at 0700 hours. The container will be closed off by ADI Magazine staff after being loaded. Once the container has been properly loaded and the M041/1&2 form signed, we will distribute that copy of the form accordingly.

The container will be padlocked, and the seals will be fixed to the container after the AMSA inspection at Patricks - No.4 North Darling Harbour on Thursday, 24 August 00 at 2000 hours - time to be confirmed.

ANZDL are requested to keep us advised of the time the full container is required at Darling Harbour.

Regards,

Stephen R. Lacy

E-mail: stephenlacy@deflog.com.au

Figure 13.2 Example notification of intent to load dangerous goods.

Source: Author's own.

IMO DANGEROUS GOODS DECLARATION

1 Shipper	2 Transportation Document Number	
	3	4 Shipper's Reference
6 Consignee	5 Freight Forwarder's Reference	
	7 Carrier (to be declared by the Carrier)	

SHIPPER'S DECLARATION
I hereby declare that the contents of this consignment are fully and accurately described above by the proper shipping name(s), and are classified, packaged, marked and labeled/placarded, and are in all respects in proper condition for transport according to applicable international and national governmental regulations.

10 Vessel/Flight and Date	9 Additional Handling Information
11 Port/Place Handling	

14 Shipping Marks	Number and Kind of Packages, Description of Goods	GW (kg)	CUBE (m3)

15 CTU ID No.	16 Seal No.	17 CTU Size and Type	18 Tare Mass (kg)	19 Total Gross Mass (kg)

CONTAINER/VEHICLE PACKING CERTIFICAT I hereby declare that the goods described above have been packed/loaded into the container/vehicle identified above in accordance with the applicable provisions.	21 Receiving Organization Receipt Received the above number of packages/containers/trailers in apparent good order and condition, unless stated hereon: RECEIVING ORGANIZATION REMARKS:	
20 Name of Company	Hauler's Name	22 Name of Company Preparing Note
Name/Status of Declarant	Vehicle Registration No.	Name/Status of Declarant
Place and Date	Driver Name and Date	Place and Date
Signature of Declarant	Driver's Signature	Signature of Declarant

Figure 13.3 Example IMO dangerous goods declaration.

Source: Author's own.

OOCL

We take it personally

Shipper Certificate and Container Packing Certificate

QF004
OHKI-031.97

# (booking approval no.)	OOLU994222005...	Page _____ of _____	
Emergency contact party	WILHELMSEN LINES A/S	& Phone no: _____	
Emergency Response Guide No	2 - 13	MFAG Number	620
Bill of Lading No.	OOLU994222005 **Vessel / Voyage**	MV KAGORO V.017S	
Shipper	MAGELLAN MARITIME LTD	Container #	OOLU3245935 (20'CP)
		(Size/Number/Type, if container known)	
Propper Shipping Name	AEROSOLS		
Hazard Class/Division (Incl. Sub-Label, if applicable)	2.1	UN Number	1950
Packaging group:	1 /11/ 111	Marine Pollutant	Yes/No

RQ *(For USA, if applicable) (for hazardous Substances)* _____ kg/ _____ lbs.
Poison inhalation Hazard   Yes/No. If Yes- then please indicate if it is a)
Poison inhalation Hazard Zone: (For USA) (A, B, C or D) _____
Dangerous When Wet   Yes/No

Quantity/ Outer package Type	310 FIBREBOARD CARTONS
Quantity/Inner Package Type	310 FIBREBOARD CARTONS
Weight (Gross/ Net)	1240 KGS. / 1000 KGS
Net Black Powder Content (kgs) (for class 1)	_____
Cubic Feet (Gross volume)	_____
Flash Point (deg C) (for class 3)	_____
Local Category (PSA CLASS):	H.K. CAT 5 (IMCO PAGE : 2101)
Placard(s)/Label(s)	AEROSOLS

Additional label (s) _____

Hazardous approval authority: _____ Hazardous approval ref. No: _____

Applicable to self-reactive Class 4.I & 5.2 Organic Peroxide (9.3.12 IMDG):	**Controlled/Set Temperature:** **Emergency Temperature:**	**deg. C** **deg. C**

CONTAINER PACKING CERTIFICATE: (Statement)

It is declared that the packing of the container has been carried out in accordance with the provisions of 12.3.7 of section 12 of the General Introductions to the IMDG Code. I hereby declare that the contents of this consignment are fully and accurately described above by the correct technical name(s) (proper shipping name(s)), and are classified, packaged, marked and labelled / placarded, and are in all respects in proper condition for transport by vessel, rail, truck according to the applicable international and national regulations.

Figure 13.4 Example shipper certificate and container packing certificate.

Source: Author's own.

Form MO41/3(a)

NOTIFICATION OF INTENTION TO HANDLE[1] DANGEROUS CARGO[2]

To the Surveyor in Charge, port of

It is intended to handle the Dangerous Cargo specified in the summary given below, the details of which are contained in the attached Form MO41/3(b)[3]

Vessel:	IMO Number:	Voyage No:
E.T.A.:	E.T.D.:	Port:
Agency:		Berth:

The handling of Dangerous Cargo is expected to commence:

Unloading:	Time: Date:	Loading:	Time: Date

SUMMARY OF CLASSES / CATEGORIES OF DANGEROUS CARGO TO BE HANDLED OR REMAIN ONBOARD

IMDG Code Class / MARPOL Category	Dangerous Cargo (yes/-) to:[4]			
	Unload	Remain Onboard	Restow	Load
1				
2				
3				
4				
5				
6				
7				
8				
9				
A				
B				
C				
D				
Approx. total mass or total no. of containers				

Signed: Print Name:	Date: Contact: Telephone Facsimile: Email:

---

[1] Includes dangerous cargo intended to remain onboard.
[2] To be submitted to the Prescribed Person not less than 48 hours prior to the intended handling. (This form may also be acceptable to the Port Authority as an Application to Handle DGs). The term Dangerous Cargo includes Harmful Substances (MARPOL Annex III) and Hazardous Wastes (Basel Convention).
[3] Form MO 41/3(b) is supplied by the Port Authority.
[4] Write "yes" in the relevant cell if any quantity of class 1, 2.1, 2.3 or 7 is onboard, to be loaded or discharged, and if more than 5 tonnes of any other class is onboard, to be loaded or discharged. Otherwise, leave blank.

Figure 13.5 Notification of intention to handle dangerous Cargo-Form MO 41_3a.

Source: Author's own.

Form M041/3(b)

MANIFEST OF DANGEROUS CARGO TO BE HANDLED OR REMAIN ONBOARD

| Vessel: | | | | | | Voyage No: | | Port: | | | Date: | |

Proper Shipping Name	Class[1]	UN Nos[2]	MARPOL[3]	Nos and Kind of Packages[4]	Total Mass	Container I.D. Nos[5]	Stowage Position	D/T/ R/L[6]	Consignee[7] or Shipper[8]	Contact[9] Name and Phone

[1] Include Flashpoint or Flashpoint Range, and subsidiary risk if applicable.
[2] Include Packaging Group if applicable.
[3] Insert P or PP for a Marine Pollutant as designated by the IMDG Code, or Category Letters A, B, C or D as appropriate.
[4] Where more than one type of dangerous cargo is stowed in the same container, this should be listed on consecutive lines of this form.
[5] For breakbulk cargo, insert B/B. For solid cargo, insert BULK. For bulk noxious liquids, insert NLS.
[6] D = cargo to be discharged. T = cargo remaining onboard in transit. R = cargo in transit for restowage. L = cargo to be loaded.
[7] For Import cargo.
[8] For Export cargo.
[9] Of consignee or shipper, as appropriate.

Figure 13.6 Notification of intention to handle dangerous Cargo-Form MO 41_3b.

Source: Author's own.

13.8 PROPOSED LOADING PLANS AND FINAL STOWAGE PLANS

Marine Orders Part 41 states that there must be onboard before the commencement of loading a plan showing the proposed stowage location of the dangerous goods. At the completion of loading, there must be a final stowage plan showing the exact location of the dangerous goods loaded, the type, class and amount. A dangerous goods manifest must also be available onboard, and copies provided to the agents and authorities (Figure 13.7).

DANGEROUS CARGO MANIFEST

ANZDL   DANGEROUS CARGO MANIFEST

VESSEL : Direct Kea		**VOY NO** : 480N			**LOAD PORT** : Sydney			
NATIONALITY : Liberian		**CALL SIGN** :ELWN7			**DISCHARGE PORT** : Oakland			
PROPER SHIPPING NAME		**IMO CLASS**	**UN NUMBER**	**GROSS WEIGHT**	**PKG GP**	**NUMBER/TYPE OF PKGS**	**CONTAINER NUMBER**	**STOWAGE POSITION**
Powder, smokeless in fibreboard cartons		1.3C	161	10077kg		1238 units	CRXU2111911	010582
24 hr no. USA : 1800 255 9324								

PREPARER'S SIGNATURE : Manjur Khan   **MASTER'S SIGNATURE** :   Capt: B. Haftendorn

Figure 13.7 Example dangerous cargo load plan.

Source: Author's own.

13.9 PRECAUTIONS DURING THE LOADING, CARRIAGE, AND DISCHARGE OF DANGEROUS GOODS

13.9.1 General Fire Precautions to Be Taken When Carrying Dangerous Goods

The general fire precautions to be taken when carrying dangerous goods include keeping the combustible material away from ignition sources; protecting flammable substances by adequate packing; rejecting damaged or leaking packages; stowing packages protected from accidental damage or heating; segregating packages from substances liable to start or spread fire; where appropriate and practicable, stowing dangerous goods in an accessible position so that packages in the vicinity of a fire may be protected; enforcing the strict prohibition of smoking in dangerous areas and display clearly recognisable 'no smoking' notices or signs; the dangers from short-circuits, earth leakage, or sparking should be apparent; and lighting and power cables and fittings should be maintained in good condition.

13.9.2 Segregation

Segregation of dangerous goods plays an important role in reducing the risks of any incident. Substances likely to react with one another are kept away from the other by various means. Primarily the class of the dangerous good is considered when deciding upon the means of segregation. The subsidiary risk is also considered when the segregation table is consulted. Tables 13.10–13.12 state the various means of segregation.

Table 13.10 Table of Segregation of Freight Containers Onboard Container Ships

Segregation Requirement		Vertical			Horizontal					
					Closed versus Closed		Closed versus Open		Open versus Open	
		Closed versus Closed	Closed versus Open	Open versus Open	On Deck	Under Deck	On Deck	Under Deck	On Deck	Under Deck
'Away from'	Fore and aft	One on top of the other permitted	Open on top of closed permitted Otherwise as for open versus open	Not in the same vertical line unless segregated by a deck	No restriction	No restriction	No restriction	No restriction	One container space	One container space or one bulkhead
	Athwartships				No restriction	No restriction	No restriction	No restriction	One container space	One container space
'Seperated from'	Fore and aft	Not in the same vertical line unless segregated by a deck	As for open versus open		One container space	One container space or one bulkhead	One container space	One container space or one bulkhead	One container space	One bulkhead
	Athwartships				One container space	One container space	One container space	Two container spaces	Two container spaces	One bulkhead
'Separated by a complete compartment or hold from'	Fore and aft				One container space	One bulkhead	One container space	One bulkhead	Two container spaces	Two bulkheads
	Athwartships				Two container spaces	One bulkhead	Two container spaces	One bulkhead	Three container spaces	Two bulkheads

Table 13.10 (Continued)

Segregation Requirement	Vertical			Horizontal					
	Closed versus Closed	Closed versus Open	Open versus Open	Closed versus Closed		Closed versus Open		Open versus Open	
				On Deck	Under Deck	On Deck	Under Deck	On Deck	Under Deck
'Separated longitudinally by an intervening complete compartment or hold from'	Prohibited			Fore and aft					
				Minimum horizontal distance of 24 metres	One bulkhead and minimum horizontal distance of 24 metres1	Minimum horizontal distance of 24 metres	Two bulkheads	Minimum horizontal distance of 24 metres	Two bulkheads
				Athwartships					
				Prohibited	Prohibited	Prohibited	Prohibited	Prohibited	Prohibited

1 Containers not less than 6 metres from intervening bulkhead.

All bulkheads and decks should be resistant to fire and liquids.

Table 13.11 Table of Segregation for IMO Dangerous Goods

CLASS		1.1 1.2 1.5	1.3 1.6	1.4	2.1	2.2	2.3	3	4.1	4.2	4.3	5.1	5.2	6.1	6.2	7	8	9
Explosives	1.1 1.2 1.5				4	2	2	4	4	4	4	4	4	2	4	2	4	X
Explosives	1.3 1.6				4	2	2	4	3	3	4	4	4	2	4	2	2	X
Explosives	1.4				2	1	1	2	2	2	2	2	2	X	4	2	2	X
Flammable gases	2.1	4	4	2	X	X	X	2	1	2	X	2	2	X	4	2	1	X
Non-toxic, non-flammable gases	2.2	2	2	1	X	X	X	1	X	1	X	X	1	X	2	1	X	X
Toxic gases	2.3	2	2	1	X	X	X	2	X	2	X	X	2	X	2	1	X	X
Flammable liquids	3	4	4	2	2	1	2	X	X	2	1	2	2	X	3	2	X	X
Flammable solids (including self-reactive and related substances and desensitised explosives)	4.1	4	3	2	1	X	X	X	X	1	X	1	2	X	3	2	1	X
Substances liable to spontaneous combustion	4.2	4	3	2	2	1	2	2	1	X	1	2	2	1	3	2	1	X
Substances, which, in contact with water, emit flammable gases	4.3	4	4	2	X	X	X	1	X	1	X	2	2	X	2	2	1	X
Oxidising (substances) agents	5.1	4	4	2	2	X	X	2	1	2	2	X	2	1	3	1	2	X
Organic peroxides	5.2	4	4	2	2	1	2	2	2	2	2	2	X	1	3	2	2	X
Toxic substances	6.1	2	2	X	X	X	X	X	X	1	X	1	1	X	1	X	X	X
Infectious substances	6.2	4	4	4	4	2	2	3	3	3	2	3	3	1	X	3	3	X
Radioactive material	7	2	2	2	2	1	1	2	2	2	2	1	2	X	3	X	2	X
Corrosive substances	8	4	2	2	1	X	X	X	1	1	1	2	2	X	3	2	X	X
Miscellaneous dangerous substances and articles	9	X	X	X	X	X	X	X	X	X	X	X	X	X	X	X	X	X

Numbers and symbols relate to the following terms as defined in Part 7: Provisions Concerning Transport Operations of the IMDG Code:

1 – 'Away from'
2 – 'Separated from'
3 – 'Separated by a complete compartment or hold from'
4 – 'Separated longitudinally by an intervening complete compartment or hold from'
X – The segregation, if any, is shown in the Dangerous Goods List
· – Refer to subsection 7.2.7.2 of Part 7: Provisions Concerning Transport Operations of the IMDG Code

As the properties of substances, materials, or articles within each class of dangerous goods may vary greatly, the dangerous goods list should always be consulted for particular provisions for segregation as, in the case of conflicting provisions, these take precedence over the general provisions.

Segregation should also take account of a single subsidiary risk label.

Table 13.12 Table of Segregation for Bulk Materials possessing Chemical Hazards and Dangerous Goods in Packaged Form

Bulk materials (classified as dangerous goods)	Class	1.1 1.2 1.5	1.3 1.6	1.4	2.1	2.2	2.3	3	4.1	4.2	4.3	5.1	5.2	6.1	6.2	7	8	9
Flammable solids (including self-reactive and related substances and desensitised explosives)	4.1	4	3	2	2	2	2	2	X	1	X	1	2	X	3	2	1	X
Substances liable to spontaneous combustion	4.2	4	3	2	2	2	2	2	1	X	1	2	2	1	3	2	1	X
Substances, which, in contact with water, emit flammable gases	4.3	4	4	2	1	X	X	2	X	1	X	2	2	X	2	2	1	X
Oxidising (substances) agents	5.1	4	4	2	2	X	X	2	2	2	2	X	X	1	3	1	2	X
Toxic substances	6.1	2	2	X	X	X	X	X	X	1	X	1	1	X	1	X	X	X
Radioactive material	7	2	2	2	2	2	2	2	2	2	2	–	2	X	3	X	2	X
Corrosives	8	4	2	2	1	X	X	1	2	2	2	2	2	X	3	X	X	X
Miscellaneous dangerous substances and articles	9	X	X	X	X	X	X	X	X	X	X	X	X	X	X	X	X	X
Materials hazardous only in bulk (MHB)		X	X	X	X	X	X	X	X	X	X	X	X	X	X	X	X	X

Numbers and symbols relate to the following terms as defined in Part 7: Provisions Concerning Transport Operations of the IMDG Code:

1 – 'Away from'
2 – 'Separated from'
3 – 'Separated by a complete compartment or hold from'
4 – 'Separated longitudinally by an intervening complete compartment or hold from'
X – The segregation, if any, is shown in the Dangerous Goods List or the individual entries in the Code of Safe Practice for Solid Bulk Cargoes

Unless other specified by the IMDG Code or in the Dangerous Goods List, segregation between bulk materials possessing chemical hazards and dangerous goods in packaged form should be in accordance with this table.

13.9.3 Packing

Packing must be well made and in good condition. Interior of packing must not be affected by contents. Packing must be strong enough to withstand normal handling. Where absorbent or cushioning material is used, such material must be capable of minimising the dangers to which the liquid may give rise, so disposed as to prevent movement and ensure the receptacle remains surrounded, and if reasonably possible, capable of absorbing liquid in the event of breakage. Receptacles of dangerous liquids must have sufficient ullage to allow for expansion to the highest temperature to be experienced during normal carriage. Cylinders or receptacles for gas under pressure must be adequately constructed, tested, maintained, and correctly filled. Empty receptacles used previously for carriage of dangerous goods must be treated as full until thoroughly cleaned.

In this chapter, we have discussed the main principles regarding the safe handling, stowage, and carriage of IMDG cargoes. Because of their intrinsically hazardous nature, IMDG cargoes require special handling and constant attention. The regulations regarding their stowage and carriage are extensive and must be followed without exception. In the next chapter, we will look at the duties and responsibilities of the cargo officer.

Part 3

Cargo Watches in Port and Officer of the Watch Responsibilities

LEARNING OUTCOME 3

On completion of this subject, you should be able to perform port watch-keeping duties.

ASSESSMENT CRITERIA

3.1 The duties of watchkeeper with regard to cargo operations are described.

3.2 The hatch and cargo gear preparations that should be carried out prior to the commencement of cargo operations are explained.

3.3 The safety checks that should be carried out prior to the commencement of cargo operations are stated.

3.4 The watchkeeper's responsibilities with regard to crew, stevedores, and visitors are explained.

3.5 The recommended safety procedures detailed in Marine Order Part 28 and STCW Code Section A-VIII/2 Part 4 that should be followed in port by the watchkeeper with regard to the following are outlined:
   - Anchors and moorings.
   - Boarding equipment.
   - Bunkering.
   - Stores and provisions.
   - Freshwater.
   - Flags and signals.
   - Garbage disposal.
   - Repairs and maintenance.
   - Lighting.
   - Stowaways.
   - Theft and pilferage.
   - Port and local regulations.

DOI: 10.1201/9781003354338-16

3.6   Correct procedure for taking over, keeping, and handing over the port watch is explained.

3.7   The importance of establishing and maintaining effective communication with personnel concerned with cargo operations is discussed.

3.8   The features of cargo stowage plans are explained and interpreted.

3.9   A pre-loading plan and a cargo plan is prepared.

3.10 Lashing and rigging plans are prepared.

3.11 The capacity plan is interpreted.

3.12 Importance of record keeping is discussed, and entries made into the deck watchkeeper's rough notebook.

3.13 A damage report is prepared.

3.14 Entries pertaining to weather, and cargo and other operations in port are made in the deck logbook.

3.15 The salient features of the following documents are outlined:
   • Shipping notes.
   • Booking lists.
   • Boat notes.
   • Mates receipts.
   • Bills of lading.
   • Charter parties.
   • Cargo manifest.
   • Notice of intention to load dangerous goods.
   • Damage reports.
   • Note of protest.
   • Shipper's declaration.
   • Certificate of moisture content.
   • Hold cleanliness certificate.
   • Certificate of fumigation.
   • Empty hold certificate or certificate of discharge.
   • Temperature control sheets.
   • Register of materials-handling equipment.

Chapter 14

Duties and Responsibilities with respect to Cargo Operations

14.1 INTRODUCTION

The OOW is primarily the master's representative when on duty on deck. Their task revolves around making and keeping a safe environment for the crew, the ship, and its cargo, and for taking measures to prevent any form of pollution from contaminating the marine environment. The OOW is helped in achieving this by guidance from their senior officers, standing orders, and legislation. By far, the most important legislation is the *International Convention on Standards of Training, Certification and Watchkeeping for Seafarers, 1978* or STCW Code. During cargo operations, the OOW must always keep in mind the safety of personnel. The safety of people, the vessel, the environment, and the cargo must be top priorities. Some steps that must be taken to ensure the safety of personnel are measures to prevent any leakages on deck or under deck; the location of life-saving appliances (LSA) and firefighting appliances (FFA); the state of readiness of LSA/FFAs; the maintenance of safe atmospheres throughout the vessel; access to and from within the vessel; adequate lighting in the holds, on deck, and all other spaces as required; being aware of the people onboard and their locations; ventilating enclosed spaces before entering; advising crew members of the dangers involved in mishandling dangerous goods; ensuring the correct rigging of cargo gear; ensuring the safe working limit of cargo gear is not exceeded; and for checking the condition of the cargo gear being used.

This list is not exhaustive and should be amended by the responsible officer as the circumstances dictate. Safety of people onboard or in the surrounding environment must never be compromised. The safety of the ship and cargo can be ensured by taking the following measures:

- Checking the draught and under keel clearance (ukc) of the vessel at regular intervals.
- Ensuring there is no excessive trim or list.
- Checking the moorings regularly.

- Checking the bilges and ballast conditions.
- Checking the tidal conditions and weather likely to be encountered.
- Understanding the cargo loading plan and the master's (and/or chief officer's) instructions.
- Knowing the various communication channels with shore authorities, Port State Control, etc.
- Understanding the causes of cargo damage and provisions for damage prevention.

14.2 DUTIES OF THE OFFICER OF THE WATCH IN THE PORT OF LOADING

The OOW has multifaceted duties and responsibilities, but when in the port of loading, they must assume additional duties required for the safe and efficient loading and discharge of cargo. These typically include, but are by no means limited to, studying and understanding the loading and de-ballasting programme; ensuring all hatch covers are secured in position, whether open or shut, and cannot move by accident; monitoring the position of the loading arm, the loading sequence and the loading rate; ensuring that the correct cargo is loaded, and the cargo is in good condition; keeping any list within acceptable limits; watching the draught to ensure that no overloading occurs; monitoring the de-ballasting to ensure that the best de-ballasting rate is maintained, any problems are identified and corrected, and maximum discharge of ballast is achieved; coordinating and supervising the work of the crew to ensure efficiency in shifting of the ship, preparing ballast holds for loading, ensuring that ship's cargo gear is properly used and maintained in perfect condition, and securing of holds on completion of cargo operations; keeping the loading foreman informed of any developments, particularly of potential problems which may affect the loading operation; noting any possible causes of damage to the ship or cargo, and making every effort to prevent them; noting and recording any damage to the ship or cargo, and immediately passing details to the chief officer who will hold the stevedores responsible; ensuring moorings and means of access are tended as necessary; preventing pollution from ballast, bilges, leakage of oil, garbage, cargo residues, funnel smoke and dust from cargo; recording full weather observations at least three times daily; maintaining full written records in the port logbook and/or deck logbook of all relevant events and data; and ensuring that safe working procedures are always followed.

14.3 DUTIES OF THE CHIEF OFFICER IN THE PORT OF LOADING

Like the OOW, the chief officer also has specific duties and responsibilities when the vessel is in the port of loading. These duties typically include, but are not limited to, the following: keeping the master fully informed of progress in the loading and of any problems encountered; distributing copies of the loading/de-ballasting plan to the loading foreman and the OOW and ensuring that it is fully understood; providing the OOW additional written instructions regarding the loading operation if the circumstances require it; keeping the loading foreman informed regarding requirements for trimming, and possible causes of delay; conducting ship's draught surveys or undertaking draught surveys with an independent surveyor, when appointed; monitoring the commencement of loading and acting promptly to deal with any problems that might arise; using informal draught surveys to monitor the tonnages delivered from time to time during loading; supervising the final stripping of ballast tanks to ensure minimum ballast is retained; calculating and supervising the trimming pours; supervising the trimming of holds filled with low-density cargo to ensure that no space is lost; supervising the trimming of holds when a level stow on completion has been stipulated; checking the space remaining in part-filled holds for the calculation of stowage factors; ensuring that appropriate matters receive attention when cargoes are loaded; providing verbal warnings, which are quickly followed by written notice, to stevedores when the ship or the cargo is damaged; ensuring the ship is properly secured for sea; and ensuring that safe working procedures are always followed. Additionally, if at anchor, the chief officer must also ascertain the vessel's position regularly; maintain a proper look out for possible dangers such as small boats or floating objects such as logs; and check the anchor specially at the turn of tide.

Irrespective of the chief officer's duties, the OOW must always be in control of the situation. The OOW has enormous responsibilities towards all the people onboard and should not hesitate to use their authority when necessary.

14.4 HATCH AND CARGO GEAR PREPARATIONS

Cargo hold preparations were dealt with in a previous section. As for cargo gear preparations, the following is an outline of an approach to the task. The use of shore cargo gear does not relieve the OOW from their duty to maintain a safe working environment. The OOW must ensure that the gear used is not likely to cause any accident. This could entail a visual inspection being used by the shore labour. Additionally, should the OOW believe that a particular item of equipment may damage the cargo, they should raise their doubts immediately and find an alternative. When using ship's gear, the rigging must be done in accordance with the rigging plan. SWL limits must be always observed. The use of experienced winch men must not be overlooked. No undue stress is to be applied to the cargo gear. Certificates must be up to date. Above all, the gear must be inspected before use. The officer should know as to when an item is unsafe for use, as stipulated in Maritime Order 42 Appendix 6, an extract of which is provided herein.

Appendix 6: Safe Use of Materials-Handling Equipment

Maximum permissible load

1.1 Except when under test, and subject to 1.2 and 1.3, an article of materials-handling equipment must not be subjected to a load greater than its SWL.

1.2 When a single sheave block is rigged as a double whip or gun tackle, so that the load is suspended from its head fitting, the load which may be lifted is twice the SWL marked on the block.

1.3 A crane or derrick may be used to hoist a load exceeding the SWL of the crane or derrick as an occasional lift, not in the course of normal operations, provided:

(a) the crane or derrick has a SWL not more than 50 tonnes;

(b) the crane or derrick has been inspected by a responsible person who is satisfied that the crane or derrick and its associated equipment is fit to carry the excess load;

(c) written permission of the master or owner in the case of ship equipment or the owner in the case of shore equipment has been obtained;

(d) the prescribed person has approved the handling of that occasional lift on a report by a surveyor; and

(e) the load does not exceed the proof load for the crane or derrick gear.

1.4 In the case of equipment with a SWL of 50 tonnes or more, 1.3(b) to (e) must be complied with and, in addition:

(a) the crane or derrick must be classed;

(b) the classification society must concur with the overloading; and

(c) the method of loading must be such that the safety of the ship and persons on it would not be imperilled by breakage of any part of the equipment, including purchase or topping lift wires.

Unsafe factors. An article of materials-handling equipment must not be rigged, reeved or used:

(a) in such a manner or under such conditions as to involve risk of injury to persons or

(b) damage to property;

(c) if the article is in such deteriorated condition or is so damaged that it may be unsafe to use; or

(d) otherwise than in accordance with this Part.

Suspended load. A load, other than, for example, a spreader or cargo lifting beam, must not be left suspended from, or supported by, a derrick, crane or mechanical stowing appliance unless, during the time it is suspended or supported, a qualified person is at the control position of the equipment engaged in the operation.

Cargo space lookout. Where persons are in a cargo space in connection with loading or unloading, whether or not a crane or derrick is being used, there must be a lookout who.

(a) has a good view of the space;
(b) is able to see potential dangers to the persons in the space; and
(c) is able to communicate with the persons in the space, and who must warn persons in the space of any perceived danger.

Note: The cargo space lookout may be a person with other duties, such as a hatchman or the crane driver, provided that the person is capable of performing all assigned duties effectively.

Securing of shackles. A crane or derrick gear that is ship equipment, must not be used in loading or unloading unless shackles and other similar devices to be used with the crane or derrick gear that are situated aloft and are not readily accessible, are effectively secured against accidental dislodgement or release.

Dragging of a load. A load must not be dragged by means of a runner leading from a derrick or a crane if there is a risk that the SWL of any component of the derrick, crane or associated cargo gear would be exceeded.

Note: Risk is considered to exist:

(a) when the lead would be direct from the derrick head or jib of the crane except when the nature of the cargo, its location and the nature of the surface over which it is to be dragged are such as to cause no likelihood of fouling of the load.
(b) when the load is more than one third of the SWL of the derrick or crane unless measures are taken to prevent overloading or unless the particular method has been shown to be safe.

Hoisting or lowering a person

Except in the case of access to a mobile offshore drilling unit or for the removal of an injured person from a cargo space, a person must not be hoisted or lowered in the course of cargo operations by means of a crane or derrick other than in a personnel cradle.

Use of wrought iron

An article of materials-handling equipment must not be used in loading or unloading if any part of that article that would support the load, either directly or indirectly, is made of wrought iron.

Use of grabs

A grab intended for use in loading or unloading bulk cargoes and which is to be attached to a ship's crane or derrick must be:

(a) permanently marked with its tare mass, cubic capacity and SWL;
(b) suitable for the material to be loaded or unloaded; and
(c) fit for use.

When using derricks rigged in Union Purchase mode, the angle between the runners must be kept below 120 degrees. When the angle between the runners exceeds 120 degrees, the load on each runner begins to exceed the weight of the lift, with the danger of failure of the lifting gear. Sheaves, blocks and all parts of the running rigging must be running smoothly. Shackles, pins and bolts must be properly attached and prevented from becoming loose. When operating hatch covers, an inspection around the hatch for people who may be unaware of any forthcoming operations must be done. Too many accidents have occurred whereby people have been crushed when hatches were being opened. Also, an inspection of the track will ensure that there are no obstructions that may prevent the opening of the hatch cover. When handling cross beams that are usually stowed on deck, never underestimate their momentum when swinging. If operating near void spaces (on partly opened hatch covers or in tween decks) always wear a safety harness. The use of heavy lift cranes/derricks requires expertise and, therefore, needs to be carried out very carefully. The equipment must be prepared for use in accordance with the manufacturer's instructions. Sometimes it might be necessary to stop other operations when lifting heavy lifts. All the components of the equipment must be carefully checked prior to usage. Wastage, lack of lubrication, and deformation are some of the defects that must not be tolerated. When using cranes, oil leaks will not only be a pollution risk, but can cause damage to the crane motor. Excessive list during cargo operations can render hydraulic cranes inoperative. On tankers, particular attention must be paid to crossover valves. Oil tight seals must be in good condition and the atmosphere in the pump room must be safe. Proper communication must be established with shore and emergency procedures fully understood. In the event of using ship's gear, for example, self-dischargers, the condition of the buckets and the conveyor belt must be ascertained.

14.5 SAFETY CHECKS TO BE CARRIED OUT BEFORE STARTING CARGO OPERATIONS

In addition to the measures to be taken as outlined earlier, the OOW is required to check the cargo compartment for its suitability; check the state of readiness of all LSA and FFA equipment; ensure the suitability of the equipment to be used; take measures to prevent pollution; familiarise themselves with any special port regulations, for example, scuppers plugged, non-discharge of ballast sediment, etc.; ensure that the stevedore foreman/supervisor understand the vessel's requirements; if necessary, have a towing wire rigged; appreciate the effect of cargo operations on the vessel's draught, trim, list, mooring arrangements, and gangway among others; check for safe atmospheres in enclosed spaces and pump rooms; ensure that measures to prevent the build-up of static electricity have been taken; check that proper communication channels have been established; and check that access to holds, compartments, cranes, and winch areas are safe.

14.6 OFFICER OF THE WATCH RESPONSIBILITIES WITH RESPECT TO THE SHIP'S STAFF, STEVEDORES, AND VISITORS

The OOW must ensure that the safety of all people onboard, including visitors, is at all times safeguarded. This might include taking measures such as posting of adequate lighting or preventing access to areas which are potentially dangerous.

14.6.1 Crew and Vessel

The safety of the crew can also be affected if the vessel is unsafe. Dangerous list or trim may have consequences on people's safety. The OOW must be in control of the situation. This also applies to the times when cargo operations are not being carried out, as factors such as tides or weather can affect the ship's safety. The crew must be briefed in the event of any unusual operation such as handling of heavy lifts or dangerous goods.

14.6.2 Visitors

The presence of visitors onboard calls for additional attention as often they might not be totally conversant with shipboard operations and basic safety precautions. The OOW should be aware of not only the presence of visitors, but also their location and the reason for being onboard. Unauthorised people, or people who are likely to be troublesome in one way or another must not be allowed onboard. The OOW must always bear in mind that they are the master's representative on deck and must act as such. A responsible master will not put the vessel and crew or visitors in danger. The OOW must ensure the same standards of safety.

14.7 SAFETY PRECAUTIONS WITH RESPECT TO PORT WATCHKEEPING

Marine Order Part 28, Appendix 1, Section 4, together with the STCW Code Section A-VIII/2 Part 4, outlines the principles applying to proper watchkeeping in port. While keeping a deck watch, the OOW must pay particular attention to the condition and securing of the anchor chain and moorings specially at the turn of the tide and in berths with a large rise and fall. The boarding equipment must be properly secured, non-slippery, and well-illuminated. If bunkering operations are underway, the OOW must take all necessary precautions to prevent pollution and reduce the risk of any fire outbreak. There must be a total understanding of the procedures involved, including communications and emergency procedures. The loading of stores and provisions also calls for care, especially when these are done with the ship's lifting gear. The dangers of using inadequate gear in the absence of proper ones cannot be underestimated. Care must be exercised when loading freshwater as this vital element is easily contaminated, with possibly dire consequences when the vessel is at sea. International, national, and local regulations must be adhered to. This involves the displaying of proper flags and signals. Some ports have strict regulations regarding the disposal of garbage. These must always be adhered to. The carrying out of repairs and any maintenance work is sometimes subject to local regulations. For example, a 'hot work' permit may be required for welding or cutting of steel. Or, over side scrapping and chipping is forbidden. Local rules must be checked before undertaking any special work. Before sailing an extensive search onboard might have to be done in stowaway-prone areas. This procedure must be enhanced by efficient policing while the vessel is in the port area. The risks of theft, pilferage, and piracy in port must always be on the OOW's mind. To conclude, it could be said that the OOW while performing his/her duties on a deck watch is not only concerned with the safety of the people, vessel, environment, and cargo for a limited time period during which he/she is on duty. His/her actions at the time will have an impact on the routine of the vessel for days if not weeks after. There is no doubt that the consequences of improper port watchkeeping can last forever.

14.8 TAKING AND HANDING OVER THE PORT WATCH

The handing and taking over of a watch can have very serious consequences on the lives of people onboard, on the vessel's safety, and on the cargo operations if this is not carried out properly. During that time, the incoming officer will get acquainted with the current situation onboard. At the end of the taking over procedure, he/she will assume full control of the vessel, and therefore should know exactly what is presently going on and what is expected in the following hours. He/she should be on deck early enough to make a full appraisal of the situation and get accustomed to the prevailing conditions. The items that require the OOW attention at the time of taking over are as follows:

- The depth of water at the vessel's berth, the ship's draught, and time of high and low water.
- Securing arrangements, anchors, and the state of the main engines.
- The nature of work being performed onboard including cargo operations.

- The level of water in the bilges and ballast tanks.
- Any signals and/or shapes being exhibited.
- The number and location of people onboard.
- The state of LSA and FFA.
- Standing and special orders.
- Port regulations.
- Communications procedures.
- Procedures for notifying appropriate authorities in the event of spillage or any other circumstances of importance to the safety of the vessel, crew, cargo, and protection of the environment.

14.9 IMPORTANCE OF ESTABLISHING AND MAINTAINING EFFECTIVE COMMUNICATIONS

The success of any operation depends heavily on effective communication. This is even more important in the case of cargo operations whereby ineffective communication can result in environmental disasters or death of people. The OOW must have a full understanding of their duties and what is expected of them. The OOW must be conversant with the standing orders together with any other special orders that the master might find relevant. Since communications in various parts of the world are not always the same, the OOW must ensure that the needs and limitations of their vessel are fully explained to shore personnel, especially those involved in cargo operations. On the other hand, they should know the limitations of the shore facilities and maintain permanent contact with them. In the event of an emergency, the OOW's communications skills will be put to the test, and these might make the ultimate difference. By being familiar with their work environment and knowing the dangers involved in any operation, the OOW should be one of the key players in the attempt to diffuse the unwanted situation.

14.10 GUIDELINES FOR COMPLETING THE SHIP/SHORE SAFETY CHECKLIST

The purpose of the Ship/Shore Safety Checklist is to improve working relationships between ship and terminal, and thereby to improve the safety of operations. Misunderstandings occur and mistakes can be made when ships' officers do not understand the intentions of the terminal personnel, and the same applies when terminal personnel do not understand what the ship can and cannot safely do. Completing the checklist together is intended to help ship and terminal personnel to recognise potential problems, and to be better prepared for them.

(15) (1) *Is the depth of water at the berth, and the air draught, adequate for the cargo operations to be completed?*

The depth of water should be determined over the entire area the ship will occupy, and the terminal should be aware of the ship's maximum air draught and water draught requirements during operations. Where the loaded draught means a small under keel clearance at departure, the master should consult and confirm that the proposed departure draught is safe and suitable. The ship should be provided with all available information about density and contaminates of the water at the berth. The term air draught should be construed carefully: if the ship is in a river or an estuary, it usually refers to maximum mast height for passing under bridges, while on the berth it usually refers to the height available or required under the loader or unloaders.

(2) *Are mooring arrangements adequate for all local effects of tide, current, weather, traffic, and craft alongside?*

Due regard should be given to the need for adequate rendering arrangements. Ships should remain well0-secured in their moorings. Alongside piers or quays, ranging of the ship should be prevented by keeping mooring lines taut; attention should be given to the movement of the ship caused by tides, currents, or passing ships and by the operation in progress. Wire ropes and fibre ropes should not be used together in the same direction because of differences in their elastic properties.

(3) *In emergency, is the ship able to leave the berth at any time?*

The ship should normally be able to move under its own power at short notice, unless agreement to immobilise the ship has been reached with the terminal representative, and the port authority where applicable. In an emergency, a ship may be prevented from leaving the berth at short notice by a number of factors. These include low tide, excessive trim or draught, lack of tugs, no navigation possible at night, main engine immobilised, etc. Both the ship and the terminal should be aware if any of these factors apply, so that extra precautions can be taken if need be. The method to be used for any emergency unberthing operation should be agreed accounting for the possible risks involved. If emergency towing-off wires are required, agreement should be reached on their position and method of securing.

(4) *Is there safe access between the ship and the wharf?*

The means of access between the ship and the wharf must be safe and legal and may be provided by either ship or terminal. It should consist of an appropriate gangway or accommodation ladder with a properly fastened safety net underneath it. Access equipment must be tended, since it can be damaged as a result of changing heights and draughts; persons responsible for tending it must be agreed between the ship and terminal and recorded in the checklist. The gangway should be positioned so that it is not underneath the path of cargo being loaded or unloaded. It should be well-illuminated during darkness. A lifebuoy with a heaving line should be available onboard the ship near the gangway or accommodation ladder.

(5) *Is the agreed ship/terminal communications system operative?*
Communication should be maintained in the most efficient way between the responsible officer on duty on the ship and the responsible person ashore. The selected system of communication and the language to be used, together with the necessary telephone numbers and/or radio channels, should be recorded in the checklist.

(6) *Are the liaison contact persons during operations positively identified?*
The controlling personnel on ship and terminal must maintain an effective communication with each other and their respective supervisors. Their names, and if appropriate where they can be contacted, should be recorded in the checklist. The aim should be to prevent development of hazardous situations, but if such a situation does arise, good communication and knowing who has proper authority can be instrumental in dealing with it.

(7) *Are adequate crew onboard, and adequate staff in the terminal, for emergency?*
It is not possible or desirable to specify all conditions, but it is important that sufficient personnel should be onboard the ship, and in the terminal throughout the ship's stay, to deal with an emergency. The signals to be used in the event of an emergency arising ashore or onboard should be clearly understood by all personnel involved in cargo operations.

(8) *Have any bunkering operations been advised and agreed?*
The person onboard in charge of bunkering must be identified, together with the time, method of delivery (hose from shore, bunker barge, etc.), and the location of the bunker point onboard. Loading of bunkers should be coordinated with the cargo operation. The terminal should confirm agreement to the procedure.

(9) *Have any intended repairs to wharf or ship whilst alongside been advised and agreed?*
Hot work, involving welding, burning, or use of naked flame, whether on the ship or the wharf may require a hot work permit. Work on deck which could interfere with cargo work will need to be coordinated. In the case of combination carriers, a gas-free certificate (including for pipelines and pumps) will be necessary, issued by a shore chemist approved by the terminal or port authority.

(10) *Has a procedure for reporting and recording damage from cargo operations been agreed?*
Operational damage can be expected in a harsh trade. To avoid conflict, a procedure must be agreed, before cargo operations

commence, to record such damage. An accumulation of small items of damage to steel work can cause significant loss of strength for the ship, so it is essential that damage is noted, to allow prompt repair.

(11) *Has the ship been provided with copies of port and terminal regulations, including safety and pollution requirements and details of emergency services?*
Although much information will normally be provided by a ship's agent, a fact sheet containing this information should be passed to the ship on arrival and should include any local regulations controlling the discharge of ballast water and hold washings.

(12) *Has the shipper provided the master with the properties of the cargo in accordance with the requirements of chapter VI of SOLAS?*
The shipper should pass to the master, for example, the grade of cargo, particle size, quantity to be loaded, stowage factor, and cargo moisture content. The *IMO Bulk Carrier Code* gives guidance on this. The ship should be advised of any material which may contaminate or react with the planned cargo, and the ship should ensure that the holds are free of such material.

(13) *Is the atmosphere safe in holds and enclosed spaces to which access may be required, have fumigated cargoes been identified, and has the need for monitoring of atmosphere been agreed by ship and terminal?*
Rusting of steelwork or the characteristics of a cargo may cause a hazardous atmosphere to develop. Consideration should be given to oxygen depletion in holds; the effect of fumigation either of cargo to be discharged, or of cargo in a silo before loading from where gas can be swept onboard along with the cargo with no warning to the ship; and leakage of gases, whether poisonous or explosive, from adjacent holds or other spaces.

(14) *Have the cargo-handling capacity and any limits of travel for each loader/unloader been passed to the ship/terminal?*
The number of loaders or unloaders to be used should be agreed, and their capabilities understood by both parties. The agreed maximum transfer rate for each loader/unloader should be recorded in the checklist. Limits of travel of loading or unloading equipment should be indicated. This is essential information when planning cargo operations in berths where a ship must be shifted from one position to another due to loading. Gear should always be checked for faults and that it is clear of contaminates from previous cargoes. The accuracy of weighing devices should be ascertained frequently.

(15) *Has a cargo loading and unloading plan been calculated for all stages of loading/de-ballasting or unloading/ballasting?*
Where possible, the ship should prepare the plan before arrival. To permit her to do so, the terminal should provide whatever information the ship requests for planning purposes. On ships

which require longitudinal strength calculations, the plan should take account of any permissible maxima for bending moments and shear forces. The plan should be agreed with the terminal and a copy passed over for use by terminal staff. All watch officers onboard and terminal supervisors should have access to a copy. No deviation from the plan should be allowed without agreement of the master. According to SOLAS regulation VI17, it is required to lodge a copy of the plan with the appropriate authority of the port State. The person receiving the plan should be recorded in the checklist.

(16) *Have the holds to be worked been clearly identified in the loading or unloading plan, showing the sequence of work, and the grade and tonnage of cargo to be transferred each time the hold is worked?*
The necessary information should be provided in the form as set out in Appendix 2 of the Code.

(17) *Has the need for trimming of cargo in the holds been discussed, and the method and extent been agreed?*
A well-known method is spout trimming, and this can usually achieve a satisfactory result. Other methods use bulldozers, front-end loaders, deflector blades, trimming machines, or even manual trimming. The extent of trimming will depend upon the nature of the cargo and must be in accordance with the BC Code.

(18) *Do both ship and terminal understand and accept that if the ballast programme becomes out of step with the cargo operations, it will be necessary to suspend cargo operations until the ballast operation has caught up?*
All parties will prefer to load or discharge the cargo without stops if possible. However, if the cargo or ballast programmes are out of step, a stop to cargo handling must be ordered by the master and accepted by the terminal to avoid the possibility of inadvertently over stressing the ship's structure. A cargo operations plan will often indicate cargo checkpoints when conditions will also allow confirmation that the cargo and ballast-handling operations are in alignment. If the maximum rate at which the ship can safely accept the cargo is less than the cargo-handling capacity of the terminal, it may be necessary to negotiate pauses in the cargo transfer programme or for the terminal to operate equipment at less than the maximum capacity. In areas where extremely cold weather is likely, the potential for frozen ballast or ballast lines should be recognised.

(19) *Have the intended procedures for removing cargo residues lodged in the holds while unloading been explained to the ship and accepted?*
The use of bulldozers, front-end loaders, or pneumatic/hydraulic hammers to shake material loose should be undertaken with care, as wrong procedures can damage or distort ships' steel work.

Prior agreement to the need and method intended, together with adequate supervision of operators, will avoid subsequent claims or weakening of the ship's structure.

(20) *Have the procedures to adjust the final trim of the loading ship been decided and agreed?*

Any tonnages proposed at the commencement of loading for adjusting the trim of the ship can only be provisional, and too much importance should not be attached to them. The significance lies in ensuring that the requirement is not overlooked or ignored. The actual quantities and positions to be used to achieve final ship's trim will depend upon the draught readings taken immediately beforehand. The ship should be informed of the tonnage on the conveyor system since that quantity may be large and must still be loaded when the order 'stop loading' is given. This figure should be recorded in the checklist.

(21) *Has the terminal been advised of the time required for the ship to prepare for sea, on completion of cargo work?*

The procedure of securing for sea remains as important as it ever was and should not be skimped. Hatches should be progressively secured on completion so that only one or two remain to be closed after cargo work is finished. Modern deep-water terminals for large ships may have very short passages before the open sea is encountered. The time needed to secure, therefore, may vary between day and night, summer and winter, fine weather and foul weather. Early advice must be given to the terminal if any extension of time is necessary.

14.11 SUMMARY

This chapter has dealt primarily with the duties of the OOW while on watch in port and during cargo watches. It should be borne in mind that no two persons react the same when confronting a similar situation. The individuality of the OOW is a factor that cannot be ignored. However, guidelines exist, and all officers should make use of them. A clear understanding of the recommendations under Section A of the STCW Code, the master's standing orders and Marine Orders Part 28 would be the starting points to that task.

Chapter 15

Cargo Planning, Record, and Log Keeping

To perform their duty as a cargo officer in an efficient way, the OOW must understand the cargo plan and the information that it contains. They should be able to complete all the preparations necessary before the cargo is loaded. This entails the preparation of the cargo plan, consulting the lashing and rigging plans, extracting information from the capacity plan, and so forth. Once the loading operation is complete, they may be required to write a damage report or clause a Mate's Receipt. Thus, an understanding of the documentation that accompanies any cargo loading or discharging operation is critical. Some the common terms that cargo officers will use include the *booking list*. This is a compilation of the cargo that the vessel intends to load. It is sent to the vessel by the ship's agent or ship's operator.

- *Pre-stowage plan*: The pre-stowage plan is a document which shows the proposed location of where the cargo will be stowed onboard the vessel. This plan is usually amended as cargo operations progress. The final stowage plan shows the actual location of the cargo after it has been stowed onboard the vessel. This plan is made only after all the cargo has been loaded and discharged accordingly.
- *Optional cargo*: Optional cargo is cargo that can be discharged at either one or more ports along the vessel's standard routing. Thus, it must be accessible for discharge at any of the ports which are preselected.
- *Overcarried cargo*: This means cargo that has not been discharged at the nominated port of discharge. This could happen because of inconsistency in tallying, an absence of documentation at the discharge port, or overlooking by the stevedores and/or ship's staff.
- *Bay plan*. This refers to the stowage plan of containers as they are onboard the vessel.

Refer to Chapter 1 for container ship plans. The information contained in the stowage plan shows the amount of cargo in each compartment, the type of cargo and its destination to the ship's officer, the cargo planner, and the stevedores. Other information available on the cargo plan are the loading ports, discharging ports, vessel's draught and voyage number, dangerous goods class number in the case of dangerous goods, and sometimes the carrying temperature of some special cargo. Usually, the cargo plan will also provide for a tonnage breakdown per hatch. For clarity sake, cargo plans for general cargo ships normally show cargo stowed in the lower holds, tween decks, and above deck. This is not the case for bulk carriers or tankers. For RORO ships, different plan views show the cargo stowage on the different decks. Examples of various stow plans for general cargo ships are included in Figure 4.14.

15.1 PRE-LOADING AND CARGO PLAN

To prepare the pre-stowage plan, certain items of information must be made available to the ship's officer. These include the port of loading, the port of discharge, types of cargo, amounts of cargo, the stowage factor of cargo, the broken stowage of cargo, vessel's next ports of call (in sequential order), carrying temperature of some special types of cargo, any segregation that may be necessary, classes of dangerous goods (if shipping dangerous cargo), the maximum draught permissible at each discharge port, and the size of oddly shaped and oversized cargo. This list is not exhaustive as special instructions will accompany each special cargo manifest. The cargo officer will then draw other information from various documents onboard before attending to the task of making the pre-stowage plan. These documents will be as follows:

- The capacity plan of the vessel.
- The rigging plan of the vessel.
- The stability booklet of the vessel.
- The cargo securing manual of the vessel.
- Previous cargo plans.

At this stage, several variables must be considered before the allocation of cargo to each compartment:

- Ports of call in the right order.
- Deck/tank top stresses.
- Special carrying requirement of the cargo.
- Availability of cargo-handling equipment.
- Capacity of ballast pumps.
- Availability of securing equipment.
- Draught and trim of vessel at each stage of the voyage.
- Segregation that may be required.

Again, this list is not exhaustive and forms only a guide as to a general approach to preparing the cargo plan. Now that these factors have been considered, the weights and volumes of the cargo are distributed throughout the various compartments. To do this, use is made of the information provided in the booking list. A perfect distribution would see the vessel with an adequate trim and no list, with no overstressing of any part of the ship's structure, and cargo distributed evenly in all compartments to allow for speedy discharge and proper segregation. With this pre-stowage plan in hand, the cargo officer must then make every effort not to divert from it. However, many circumstances dictate frequent amendments to the plan. This can only be done with the master's or chief officer's approval. The preparation of a pre-stowage plan involves calculations to work out the weights and volume to be occupied by the cargo in each hold. If we recall, this topic was dealt with in Chapter 1. On the completion of loading, the amended version of the pre-stowage plan is made up showing the exact location, weight, and volume of each cargo, in each compartment. This is known as the final stowage plan and is used at the discharging port to decide upon the cargo operations to be used.

15.2 CAPACITY, LASHING, AND RIGGING PLANS

Onboard the vessel, three plans are particularly important to the chief officer when it comes to cargo work:

(1) The capacity plans.
(2) The lashing plans.
(3) The rigging plans.

15.2.1 Capacity Plan

This plan is very commonly displayed in the alleyways, or in the ship's office, or very near the chief officer's office. It gives a general view of the vessel, together with the volumes of the compartments (in bales and grain capacities), the sizes of the compartments, the maximum allowable deck stresses, the stacking loads, capacity of all onboard tanks, deadweight, lightship, position of the centre of gravity of compartments, and the various dimensions of the vessel.

15.2.2 Lashing Plan

Inadequate securing on ships has resulted in loss of cargo, ships, and crew. The aim of the lashing plan is to assist the officer in determining what is the minimum securing arrangements that are required for a particular type of cargo in normal sea conditions. Each lashing plan is tailor-made for a particular vessel following observation of her behaviour at sea. The securing equipment mentioned in the plan are of a tested and approved type. It must be remembered that in heavy weather, the mariner must exercise extra care, and this may mean extra lashing arrangements. The lashing plan is made following guidelines stipulated in the IMO publication *Guidelines for the Preparation of the Cargo Securing Manual* which every vessel must now carry.

15.2.3 Rigging Plan

This plan shows the capabilities and limitations of the cargo lifting gear. Derricks with their safe working limit, minimum and maximum outreach, and different rigging arrangements are some of the features of the rigging plan. It is consulted when a derrick must be used in a different mode than the one it is normally rigged. For example, two derricks are used in the union purchase mode; if, for one operation, one derrick is required as a swinging derrick, the initial arrangement must be broken, and one derrick used. Consulting the rigging plan will remove any doubt in the officer's mind as to the safest arrangement possible. Details of the materials-handling equipment, their measurements, and safe working limit are found on that plan.

15.3 RECORD AND LOG KEEPING

As mentioned earlier, cargo damage claims are the biggest claim that shipowners face every year. Quite often, the shipowner is blamed for their 'apparent lack of care'. In many cases, it is in fact a lack of documentation that prevents the shipowner from adequately defending themselves. There are various reasons why it is important to properly document cargo operations, but the main one is most probably to avoid long legal battles over the condition of the cargo. Since carriage of cargo involves caring for the cargo, it is vital that the condition of the cargo be noted at loading and compared with its condition at discharge. Should there be a major change in condition, then the shipowner will have to explain the circumstances that caused that change. No doubt that they will have to face the consequences of any negligence from the ship's crew. The cargo officer must not only note the condition of the cargo, but also the weather experienced at the time of loading and any special occurrence at the time. Above all, should the cargo officer believe that cargo has been damaged prior to loading, they must advise the chief officer immediately and report that fact in the cargo logbook. Each entry in a cargo logbook has its importance whether it is weather related (cargo being affected by rain), time related (used in working out the time during which the vessel was on/off charter), or cargo related (used to ascertain the condition in which the cargo was loaded). In shipping, one important piece of documentation is the Bill of Lading. It is a negotiable document, which means that the cargo owner could sell the cargo mentioned on the Bill of Lading to anyone.

The production of the Bill of Lading is done following the information gathered from another document known as the Mate's Receipt. The Mate's Receipt is a document stating the goods shipped onboard, the shipper's name, the amount and type of cargo, port of destination, vessel's name, and, more importantly, the condition of the cargo at loading. As the Bill of Lading is a negotiable document, it will have less value if it has been claused. Therefore, it is very important to make all the necessary entries in the cargo logbook. They might sound trivial initially, but they come in very handy when there is a need to support one argument. Claims can be reduced to a certain extent if proper documentation is in place to prove that the shipowner and/or his/her servants have acted in good faith.

To ensure accurate records are kept, the ship's officers are strongly advised to follow a checklist, which includes all records that should be maintained onboard. Whilst every vessel will have unique requirements, the most common or standard checks and records that cargo officers should maintain include the following:

Deck logbook entries:

- Routine navigational, weather, sea state, and ship's performance data.
- Details of heaving-to, or action taken to avoid tropical storms.
- Dew point readings of cargo spaces a–d on deck.
- Ventilation of holds – times of starting and stopping, reason for stopping, ventilators used, type of ventilation, direction of ventilation, speed of fans, and hygrometer readings.
- Water, rainfall, and spray over the decks or hatches.
- Hold and hatch cover inspections included the dates and times, names of the person carrying out the inspection, the nature of the inspection, and any observations and findings from the inspection.
- Temperatures, methane, and O_2 meter readings of cargo.
- pH of bilge water.
- Pumping of bilge water, including the time, tonnage, and origin.
- Soundings – usually a full set daily, giving actual soundings.
- Testing of cargo care systems such as hold bilge pumping system, hold ventilation fans, hold CO_2 injection systems, the testing of hatch cover watertightness.
- Inspection and tightening of lashings on cargo.
- Changing of ship's ballast to comply with pollution regulations, or for purposes of draught and trim.
- Details of any in-transit fumigation.

Cargo log entries

- Surveys undertaken, with times, result, and the identity of surveyor.
- Protests made by the ship, and to the ship.
- Details of any fumigation undertaken.
- Starts, stoppages, and completions of cargo work.
- Transfers of cargo-handling equipment.
- Tanks ballasted and de-ballasted.
- Ballast valves opened and shut.
- Starts and stops of ballasting, and ballast pump readings.
- Soundings obtained.
- Cargo tonnages calculated or advised.
- Draught readings at completion of each pour during the loading, and at least twice daily during discharge.
- Details of shifting ship.
- Times of bunkering, and quantities taken.
- Weather observations.

Cargo documents

- Copies of all cargo documents issued or received.
- Authorisation to charterers or their agents to sign the Bill of Lading.

Trim, stability, and stress calculations

- Values used in calculations.
- Results obtained.
- Full details of departure condition.
- Copy of each cargo operations control form issued.
- Ship's own draught survey calculations.
- Draught survey calculations by independent surveyors, and results obtained.

Damage records

- When/where/how damage to ship or cargo occurred.
- Detailed description of damage sustained.

15.4 DAMAGE REPORT

If during cargo operations, cargo is accidentally damaged, a damage report should be made immediately. The time of the incident, the extent of the damage, the cargo damaged, factors that led to the incident must be part of that report. Witnesses to the incident should sign the document. Stevedores and port authorities might have to be advised depending on the extent of the damage. Sometimes, the vessel itself might be damaged during cargo operations. Again, a damage report should be made. The damage reports are sent to the shipowner who has a legal team which is going to do the needful to recover the costs of the damage or defend the shipowner if the latter seems to be at fault. At times, if the damage to the cargo and vessel has been extensive, an independent surveyor is called in to assess the extent of the damage. He/she should be always given full assistance.

15.5 IMPORTANT SHIPPING DOCUMENTS

To fully understand the documentation involved in the carriage of cargo, some of the main definitions used are listed in Table 15.1.

Table 15.1 Important Shipping Documents and Definitions

Shipping Note	This is a document issued by the shipper to the carrier or vessel operator stating their intention to ship cargo from one place to another. Details regarding the sender of the goods (shipper), the receiver of the goods (consignee), and the type and amount of cargo are features of the shipping note
Boat Note	The term used to describe the document that accompanies the cargo from its place of origin (shipper's factory, warehouse, or even backyard) to the port warehouse or storage place. People responsible for loading the ship check the marks on the goods against that stated on the boat note before shipping it onboard
Mate's Receipt	This is a document stating the goods shipped onboard and their condition. It is signed by the Chief Officer who attests that the cargo mentioned therein is onboard. The Mate's Receipt is used to prepare the Bill of Lading
Bill of Lading	One of the most important documents in carriage of cargo. It is a document of title, a receipt for goods shipped, and is evidence of contract between the carrier and the shipper. There are many types of Bills of Lading, some of which are not negotiable (cannot be used to sell the goods mentioned in the document), others being intermodal (used to cover different types of transportation), or even in-house Bill of Lading (a Bill of Lading prepared by a freight forwarder to individual shippers when cargo is being sent as a groupage). See Chapter 1 for an example
Charter Party	The agreement that exists between a shipowner and a charterer. All the terms of the agreement and penalties in the event of breach are stated in that document
Cargo Manifest	This is a compilation of all the cargo onboard. It is made from the Bills of Lading. It is used by various port authorities, notably Customs

Table 15.1 (Continued)

Notice of Intention to Load Dangerous Goods	As the name suggests, this is a notice given to the local authorities, for example in Australia, AMSA, as to the vessel's intention to load dangerous cargo in that port. Dangerous goods cannot be loaded if prior notice has not been given. See Marine Order Part 41 for more details regarding the shipping of dangerous goods in Australia
Note of Protest	This is a document signed by the master in the presence of a Notary Public exonerating themselves and their vessel from damages that might have occurred to cargo onboard if the vessel has experienced heavy weather. They might elect to extend the protest when the damages are assessed
Shipper's Declaration	A document from the shipper giving details of the cargo. Cargo-carrying temperature, moisture content of the cargo, and stowage factor of the cargo are important information found on a shipper's declaration. Also mentions the weight and dimensions of the cargo
Dangerous Goods Declaration	This is a document like the shipper's declaration except that it deals with dangerous goods. In this case, the UN number, class, and flashpoint of the goods must be included. The document means that the shipper has packed, marked, and labelled the packages according to the IMDG Code
Certificate of Moisture Content	This is a document provided by a testing authority following the test carried out on some types of cargo (usually bulk cargoes) to determine the moisture content of that cargo. It is important as it allows the vessel to reject the cargo should the moisture content exceed the transportable moisture limit
Hold Cleanliness Certificate	The Hold Cleanliness Certificate states the condition of the hold before loading. It is issued by local authorities after the compartment has been inspected. Sometimes required by shippers before allowing their cargo to be loaded. It is quite common on gas and product carriers
Certificate of Fumigation	This is issued by approved chemist or fumigation companies following the fumigation of compartments. Sometimes some compartments must be fumigated before loading bulk grain
Empty Hold Certificate/ Certificate of Discharge	This is generally issued by an independent marine surveyor to confirm that the stevedore has completed the cargo operation in that hold
Temperature Control Sheets	Are records of duty inspection of refrigerated holds and/or containers compiled by the ship

Index

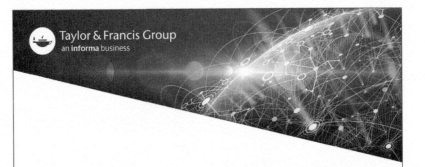